Penguin Education
Penguin Library of Physical Sciences

Gases, Liquids and Solids

D. Tabor

Advisory Editor
V. S. Griffiths

General Editors
Physics: N. Feather
Physical Chemistry: W. H. Lee
Inorganic Chemistry: A. K. Holliday
Organic Chemistry: G. H. Williams

Gases, Liquids and Solids

D. Tabor

Penguin Books

Penguin Books Ltd, Harmondsworth,
Middlesex, England
Penguin Books Inc., 7110 Ambassador Road,
Baltimore, Md 21207, U.S.A.
Penguin Books Australia Ltd, Ringwood,
Victoria, Australia

First published 1969
Reprinted 1970
Copyright © D. Tabor, 1969

Made and printed in Great Britain by
Bell & Bain Ltd, Glasgow
Set in Monotype Times

Contents

Editorial Foreword

For many years, now, the teaching of physics at the first-degree level has posed a problem of organization and selection of material of ever-increasing difficulty. From the teacher's point of view, to pay scant attention to the groundwork is patently to court disaster; from the student's, to be denied the excitement of a journey to the frontiers of knowledge is to be denied his birth-right. The remedy is not easy to come by. Certainly, the physics section of the Penguin Library of Physical Sciences does not claim to provide any ready-made solution of the problem. What it is designed to do, instead, is to bring together a small library of compact texts, written by teachers of wide experience, around which undergraduate courses of a 'modern', even of an adventurous, character may be built.

The texts are organized generally at three levels of treatment, corresponding to the three years of an honours curriculum, but there is nothing sacrosanct in this classification. Very probably, most teachers will regard all the first-year topics as obligatory in any course, but, in respect of the others, many patterns of interweaving may commend themselves, and prove equally valid in practice. The list of projected third-year titles is necessarily the longest of the three, and the invitation to discriminating choice is wider, but even here care has been taken to avoid, as far as possible, the post-graduate monograph. The series as a whole (some five first-year, six second-year and fourteen third-year titles) is directed primarily to the undergraduate; it is designed to help the teacher to resist the temptation to overload his course, either with the out-moded legacies of the nineteenth century, or with the more speculative digressions of the twentieth. It is expository, only: it does not attempt to provide either student or teacher with exercises for his tutorial classes, or with mass-produced questions for examinations. Important as this provision may be, responsibility for it must surely lie ultimately with the teacher, for he alone knows the precise needs of his students – as they change from year to year.

Within the broad framework of the series, individual authors have rightly regarded themselves as free to adopt a personal approach to the choice and presentation of subject matter. To impose a rigid conformity on a writer is to dull the impact of the written word. This general licence has been extended even to the matter of units. There is much to be said, in theory, in favour of a single system of units of measurement – and it has not been overlooked that national policy in advanced countries is moving rapidly towards uniformity under the *Système International* (S.I. units) – but fluency in the use of many systems is not to be despised: indeed, its acquisition may further, rather than retard, the physicist's education.

A general editor's foreword, almost by definition, is first written when the series for which he is responsible is more nearly complete in his imagination (or the publisher's) than as a row of books on his bookshelf. As these words are penned, that is the nature of the relevant situation: hope has inspired the present tense, in what has just been written, when the future would have been the more realistic. Optimism is the one attitude that a general editor must never disown!

N. FEATHER

January 1968

Preface

The subdivision of physics into mechanics; heat, light and sound; magnetism and electricity; properties of matter; and kinetic theory, goes back to the early days of classical physics. During the last twenty or thirty years there has been a discernible trend towards a regrouping of subjects, partly to allow for new knowledge and partly to allow for new methods of approach. Few of the areas have received such an impulse as those which, in classical days, were grouped together as kinetic theory and properties of matter. The new approach is to emphasize the atomic structure of matter and to show that by assuming the existence of attractive and repulsive forces between atoms and molecules and the presence of thermal energy, it is possible to explain nearly all the bulk properties of gases, liquids and solids in terms of relatively simple models.

The present book, which attempts to do this, is based on the lecture course given to first year students at Cambridge. Its treatment is relatively simple and contains very little quantum physics or wave mechanics. It represents an attempt to bridge the gap between sixth-form physics and physical chemistry, and the more advanced courses which follow in later years of specialization. If it has any merit at all this is largely due to the devotion of many generations of Cavendish teachers who have hammered out the type of treatment and approach I have given here. I am particularly indebted to Dr T. E. Faber who first explained the Cavendish syllabus to me, and to Dr D. Shoenberg who has run a parallel course of lectures with me for the last few years and who has, on several occasions, saved me from perpetuating mistaken ideas. I also wish to thank the members of the Cavendish who have made comments on various parts of the manuscript, and Miss Shirley V. King of Birkbeck College for her comments on the first part of Chapter 11. Any mistakes, however, either of fact or of opinion are my own and I accept responsibility for them.

Dr Leo Baeck, in his spirited critique of romantic religion, referred to orthodoxy in the following terms: 'that bent of mind which would cultivate respect for the answer but, in the process, often loses respect for the problem.' A text book such as this must naturally pay great respect to the answer. The student quite rightly expects to find the correct answer crisply and clearly given. This I have attempted to do. But it sometimes happens in physics, as in other walks of life, that the simple crisp answer is not the whole truth. Where it seems profitable I have discussed this as openly as possible. I hope it does not confuse the student, for this is not my intention.

I recognize that, so long as a student's ability is judged by examinations, both student and examiner will persist in cultivating great respect for the answer. On the other hand, the progress of physics in the long run, may well depend far more on our according greater respect to the problem.

Chapter 1
Introduction

The three states of matter are the result of a competition between thermal energy and intermolecular forces. It is this struggle which determines whether a given substance, under given conditions, is a gas, a liquid or a solid. Many textbooks which describe the bulk properties of matter tend to lose sight of this and to treat gases, liquids and solids as though they are quite unrelated materials. Indeed gases become an exercise in kinetic theory, solids an example of the laws of elasticity, and liquids a type of material which shows viscosity and surface tension. In the present book we attempt to show that many of the bulk properties of gases, liquids and solids can be explained in terms of intermolecular forces; in this way the three states are linked by a common property. We shall not give a detailed account of the origin of molecular forces but, assuming they exist, we shall show the part they play in determining the properties of matter.

Chapter 2 is a very simple account of the nature of atoms and molecules and the forces between them. The third chapter deals with temperature and the concept of heat. These are the introductory chapters.

The next three chapters describe the properties of gases. This is a well-worn area where the molecular approach has long been used to describe the bulk behaviour. Chapter 4 gives the simple kinetic theory of ideal gases, while Chapter 5 deals with a more sophisticated treatment which includes a simplified approach to the Boltzmann distribution, and a discussion of the equipartition of energy. Chapter 6 deals with imperfect gases where perfect gas behaviour is modified to allow for the finite size of the molecules and for the forces between them.

We turn at once from gases to solids. In gases the molecules are virtually free, in solids they are bound to particular sites and their main thermal exercise consists in vibrating about their equilibrium positions. However, the molecular forces that cause the real gas to deviate from ideal gas behaviour are the same forces that are responsible for the existence of the solid state. In Chapter 7 we show that these forces can explain the heat of sublimation of solids, their thermal expansion, their elastic properties and the existence of the surface energy of solids. Some of these themes are analysed in greater detail in subsequent chapters.

In Chapter 8 the elastic properties of solids are described in terms of intermolecular forces. This is applied quantitatively to ionic and van der Waals solids and empirically to metals. An account is also given of the elasticity of rubber. Chapter 9 deals with the plastic and brittle properties of solids and shows that these may be understood in molecular or atomic terms. For both types of strength property it is shown that the real strength is generally much less than the theoretical because of the presence of imperfections. Chapter 10 describes the thermal and electrical properties of solids. In this chapter we give a brief account of the specific heats of solids in terms of atomic vibrations (the Einstein and Debye theories); we also show how thermal expansion arises as the amplitude of the vibration increases with increasing temperature. Finally we describe thermal and electrical conductivity; with metals these are both attributed to the transport of energy by the free electrons. The classical theory which treats the electrons as a gas gives the right answers for the wrong reason and it is shown how the theory must be modified to allow for the discrete energy-levels of the electrons.

The next two chapters deal with the liquid state, the Cinderella of modern physics. The main difficulty here is that, although in some ways the intermolecular forces dominate (a liquid occupies a specified volume unlike a gas), in other ways thermal motion dominates (a liquid shows mobility, unlike a solid). The molecules are both bound and free so that over small regions they appear to be highly ordered as in a solid whilst over large regions such order does not exist. These features and some recent descriptive ideas (particularly those of Bernal and of Alder) are given in Chapter 11 together with a molecular theory of vapour pressure and surface tension. In Chapter 12 we discuss the flow of liquids and using the simplest form of Eyring's theory, show that viscosity may be understood in terms of intermolecular forces. Consequently the connection between surface tension and viscosity is found.

The last two chapters deal with the dielectric and magnetic properties of matter. It is shown that these properties arise from the polarization of atoms and molecules produced by the applied electrostatic or magnetic field. The dielectric behaviour, described in Chapter 13, receives a very satisfactory explanation in terms of simple atomic and molecular models. In addition the treatment provides a direct correlation between dielectric properties, optical properties and van der Waals forces. The explanation of magnetic properties is more difficult since quantum effects are of major importance. The classical explanation of diamagnetism which we reproduce in Chapter 14 yields the right answer but, at a more fundamental level, is quite wrong. Paramagnetism is more easily explained and, assuming the existence of strong interactions, a satisfactory and satisfying model for ferromagnetism is found.

Some of the chapters lend themselves to brief summaries and where this has been possible they have been introduced at or near the end of the relevant chapter. A final word on units: the system adopted throughout this book is the c.g.s. system, but conversion tables are given at the back of the book.

2 Gases, liquids and solids

Chapter 2
Atoms, molecules and the forces between them

2·1   Atoms and molecules

2·1·1   *The evidence for atoms and molecules*

Matter is not a continuum of uniform density, but consists of discrete particles, or if one wishes to be more up to date, of localized regions of very high density separated by regions of almost zero density. The particulate nature of matter has been known to us since the time of the Greeks and the idea of atoms (units which could not be further cut or divided) is generally attributed to Democritus (*c.* 460–370 B.C.). Many of his concepts have a surprisingly modern ring and constitute a tribute to the immense power, as well as the originality, of the Greek approach to logical inference and abstract reasoning. Much of his work was, indeed, the result of thought rather than of direct experiment. One must not, however, read too much modern science into these early ideas. For example in a passage quoted by Theophrastus, *De Sensu*, 61–62, Democritus states, 'Hard is what is dense, and soft what is rare . . . . Hard and soft as well as heavy and light are differentiated by the position and arrangement of the voids. Therefore iron is harder and lead heavier.' But it would be wrong from this to attribute to Democritus a knowledge of dislocations and point defects.

The first real attempt to get to grips with the basic 'atoms' of matter had to wait until more quantitative measurements and generalizations had been made. The law of multiple proportions, largely due to the work of J. Dalton (1766–1844) was one of these. It stated that if two elementary substances combined chemically to form more than one compound the weights of one which combine with a fixed weight of the other are in a simple ratio to one another. For example nitrogen and oxygen combine to form nitrous oxide, nitric oxide, nitrogen sesquioxide and nitrogen dioxide; three of these oxides were, in fact, amongst the compounds which Dalton first studied in the course of his investigation. In these compounds 14 grams of nitrogen combine respectively with 8, 16, 24 and 32 grams of oxygen: i.e. the ratio is 1 : 2 : 3 : 4. This fits in naturally with the concept of unit masses which can combine in simple multiples with one another.

We consider a simpler case, the reaction of hydrogen and oxygen to form water. Quantitatively we find 1 gram of hydrogen combines with 8 grams of oxygen to form 9 grams of water. According to Dalton's atomic theory

(Dalton 1808) we should write

$$1 \text{ atom hydrogen} + 1 \text{ atom oxygen} \rightarrow 1 \text{ atom water}$$
or
$$H + O \rightarrow HO.$$

When these reactions are studied in the gaseous state and the volumes are measured at a standard temperature and pressure, the equation becomes

$$2 \text{ volumes hydrogen} + 1 \text{ volume oxygen} \rightarrow 2 \text{ volumes water vapour.}$$

The next step in the argument depends on the hypothesis put forward by A. Avogadro in 1811 that at a specified temperature and pressure a given volume of any gas contains the same number of unit masses.

Our previous equation then becomes

$$2 \text{ unit masses of hydrogen} + 1 \text{ unit mass of oxygen} \rightarrow$$
$$2 \text{ unit masses of water.}$$

This statement makes sense only if the 'unit mass' of oxygen is itself divisible into two equal parts each of which goes into one of the unit masses of water. Arguments along these lines lead finally to chemical equations of the type

$$2H_2 + O_2 \rightarrow 2H_2O,$$

where H_2, the smallest quantity of hydrogen which can exist in the free state, is called the molecule whilst H. the smallest quantity of hydrogen to enter into chemical combination, is called the atom. Avogadro's 'unit masses' are therefore molecules.

2·1·2 *Avogadro's law*

This is still called a hypothesis but deserves to be promoted to the status of a law. It states that at a specific temperature and pressure a fixed volume of any gas or vapour contains the same number of free unit masses, i.e. molecules. If we re-define the atom of hydrogen as 1 unit mass then the molecule of hydrogen consists of 2 unit masses. These quantities in grams are the gram-atom and gram-molecule respectively (a gram-molecule of any gas at a pressure of 760 mm. Hg and a temperature of 0°C. fills approx. 22,400 c.c.). The number of molecules in a gram-molecule of any substance is called Avogadro's number N_0 and has the value of $6 \cdot 06 \times 10^{23}$. For example N_0 molecules of hydrogen weigh 2 g., of oxygen 32 g., of gold 197 g., of iodine 254 g.; these constitute the gram-molecular weights of these materials.

2·1·3 *First experimental deduction of the size of molecules*

Pioneer work by Fraülein Pockels in 1891 showed that certain organic molecules with polar end groups such as oleic acid which are insoluble in water in bulk will spread over the surface of clean water. Lord Rayleigh

4 Gases, liquids and solids

experimented further with this and showed in 1899 that if the amount of acid added to the water surface is less than a certain amount there is practically no reduction in surface tension of the water. If, however, the amount added exceeds a certain critical value the surface tension falls rapidly. This is shown schematically in figure 1. Rayleigh assumed that this change occurred

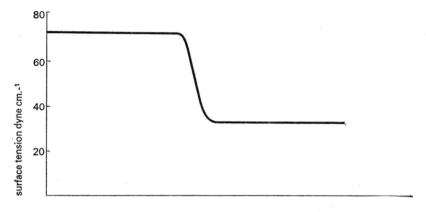

amount of surface active material added

Figure 1. Schematic diagrams showing the surface tension of water as small quantities of surface active material are added. The surface tension remains fairly high until a critical amount, corresponding approximately to a complete mono-layer of surfactant has been added; it then falls rapidly to a constant lower value.

when there was a continuous monomolecular layer of fatty acid on the surface. He was thus able to determine the weight of acid per sq. cm. of surface at which the monolayer was formed. Assuming the monolayer to have the same density as the acid in bulk the thickness of the monolayer was calculated and found to be about 10^{-7} cm., i.e. 10 Å. This was the first direct indication that atomic and molecular dimensions are of the order of angstroms.

2·1·4   *Determination of Avogadro's Number N_0*

We shall mention three very different ways of determining N_0. They are all of historic interest but only the last two provide reasonably accurate values.

2·1·4·1   *Radioactive disintegration.* Radium disintegrates to give off α particles and leave radon gas. Radon disintegrates to give off α particles and leave behind radium A. Further stages in the disintegration are not, in the present context, of great importance. The α particles emitted can be counted by allowing the particles emitted within a small known solid angle to strike a fluorescent

5   Atoms, molecules and the forces between them

screen. Each particle may be counted since it produces an individual scintillation on the screen. From the size of the solid angle chosen the total emission over a solid angle of 4π can be deduced. Lord Rutherford in 1909 showed that the α particle is a He atom which has lost its two orbital electrons (see below). These electrons are picked up by the particles so that each α particle becomes a helium atom, and since helium is a monatomic gas each α particle thus becomes a gas atom. The volume of the helium gas produced in this way can be measured. By extrapolation Rutherford found that, in one year, one gram of radium would emit a total of $11 \cdot 6 \times 10^7$ particles which would turn into $0 \cdot 043$ c.c. of helium gas at s.t.p.

> Consequently $0 \cdot 043$ c.c. He contains $11 \cdot 6 \times 10^7$ atoms.
> Therefore 22,400 c.c. He contain 6×10^{23} atoms.

This derivation is of interest since it showed for the first time that atoms are individual particles.

2·1·4·2   *Electrolysis.* If a quantity of electricity known as a faraday is used to electrolyse a dilute acid, exactly 1 gram of hydrogen is liberated. Since the faraday is equal to 96,500 coulombs, this means that 96,500 coulombs are involved in converting N_0 ions of hydrogen into N_0 atoms of hydrogen gas. The charge on the hydrogen ion is due to the loss of a single electron, hence

$$96,500 \text{ coulomb} = \text{charge of } N_0 \text{ electrons} = N_0 e.$$

The next step is to determine e. This was carried through successfully by R. Millikan in 1909. He produced very fine droplets of oil between the plates of a parallel plate condenser, the plates being set in a horizontal plane. The particles gradually fall because of gravity. If, however, the air surrounding them is subjected to ionizing radiation the particles will pick up charge of say q. It then becomes possible to apply an electrostatic field E which is able just to hold the drop in equilibrium in a stationary position. In this state

$$Eq = mg.$$

Therefore as m, g and E were known, q was calculated and found always to be an integral number of some fundamental unit of charge. This was assumed to be the charge e of the electron. The current value of e found in this way is very nearly

$$e = 4 \cdot 8 \times 10^{-10} \text{ e.s.u.} = 1 \cdot 6 \times 10^{-19} \text{ coulomb.}$$

Hence $N_0 = \dfrac{96,500}{1 \cdot 6 \times 10^{-19}} = 6 \cdot 03 \times 10^{23}.$

2·1·4·3   *X-rays.* X-rays are electromagnetic waves of short wavelength (about 10^{-8} cm.). M. von Laue was the first to suggest, in 1912, that since the spacing of crystal planes in a simple inorganic crystal is of this order, it ought to be possible to use the crystal as a diffraction grating for X-rays. The waves length λ of the X-rays may be determined accurately by diffracting them with a conventional ruled grating, at glancing incidence. The X-rays may then be used to find the lattice spacing in the crystal. If, for example, a collimated beam of X-rays strikes a crystal plane at an angle θ_1, it is first refracted through an angle θ_2, but since the refractive index for X-rays is almost unity one may take $\theta_1 = \theta_2 = \theta$. The condition for reinforcement is then that $2d \sin \theta = n\lambda$, where d is the distance between scattering planes and n is an integer. By observing the angles for which strong diffracted beams are obtained, the various scattering planes and their separation may be found. In this way the crystal structure and the lattice spacings may be deduced.

As an example consider the sodium chloride crystal. This has a cubic structure with sodium and chloride ions at alternate sites (see figure 2).

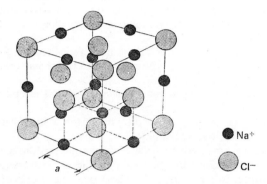

Na$^+$

Cl$^-$

Figure 2. Structure of sodium chloride crystal. The sodium and chloride ions are arranged alternately on a cubic lattice, the separation a between ions being 2·818 Å. The sodium fluoride structure is identical except that $a = 2·31$ Å. The 'unit cell' size is $2a$.

The distance a between each site is found by X-ray analysis to be 2·818 Å. Consider one of the unit cubes of side a. Each of these cubes has 4 sodium and 4 chloride ions, i.e. 4 NaCl molecules. But each corner of the cube is shared by eight contiguous cubes. Consequently each volume a^3 contains one eighth of four molecules of NaCl = $\frac{1}{2}$ molecule of NaCl. If M = molecular weight, ρ = density, the volume occupied by the half molecule is $\frac{1}{2} \frac{M}{N_0} \frac{1}{\rho}$. Consequently

$$a^3 = \frac{1}{2} \frac{M}{N_0 \rho}.$$

Inserting $M = 58·5$, $\rho = 2·17$, $a = 2·818$ Å.,

$$N_0 = 6·02 \times 10^{23}.$$

7 Atoms, molecules and the forces between them

1. An atom consists of a minute nucleus, which is positively charged and contains the main mass of the atom.

2. The nucleus consists of protons and neutrons. These are of equal mass but each proton carries a positive charge whilst the neutron carries no charge. The resultant positive charge (referred to the proton as unit charge) is called the atomic number Z, that is Z is equal to the number of protons in the nucleus.

3. As a crude approximation, valid especially for smaller atoms of atomic number less than 20, there are about equal numbers of protons and neutrons so that the mass of the nucleus, referred to the proton as unit mass, is about twice the atomic number. This is known as the mass number and in practice is the integer nearest to the mass of the atom, referred to the carbon-12 atom as 12·000. In the older literature the mass number is sometimes loosely referred to as the atomic weight.

4. The nucleus is surrounded by electrons. These have a mass only $\frac{1}{1840}$ of the mass of the H atom (i.e. of a proton or neutron) and they have a unit negative charge.

5. Since the atom is electrically neutral the number of orbital electrons is equal to the atomic number.

6. The chemical properties of the atom are determined by the configuration of the electrons, especially of the outermost ones.

7. There are certain stable arrangements, the electrons falling into well defined energy levels or shells, defined by quantum conditions. Each energy level can hold no more than two electrons and these must have opposite spins (see 8 below). This is known as the Pauli Exclusion Principle.

8. The lowest energy shell, with principal quantum number 1, is the K shell. This ground-state is called the 1s orbital. Here the electron is distributed with spherical symmetry about the nucleus and has zero angular momentum. However, the electrons can have a spin quantum number which can be described as being either parallel or antiparallel to some specified direction, for example, ↑ or ↓. If the K shell contains 1 electron as in hydrogen it has a single spin; if it contains two electrons as in helium the spins are paired in opposite directions and the K shell is now complete. The atom is thus inert.

9. The next higher energy shell has principal quantum number 2 and is known as the L shell. The lowest level is called the 2s state (again spherically symmetrical) and this can hold 2 electrons of opposite spin. There is a higher energy orbital known as the 2p. This has angular momentum, is dumb-bell in shape, and there are three directions available. Each can hold 2 electrons of paired spins so that altogether this state can hold 6 electrons. If only two or three electrons have to go into the 2p state the energy is lowest if they each have parallel spins (this is part of a more general principle, due to F. Hund, which says that when several levels within a given state are available the lowest energy corresponds to parallel spins): they can do this by each

going into the three available orbitals. These electrons can readily play a part in chemical bonding and other processes. A fourth electron will have to pair its spin with one of the existing three. The remaining two unpaired electrons are then readily available for chemical reaction. When the $2p$ state is filled with its six electrons the atom – neon – is chemically non-reactive (but see below).

10. The next higher level has principal quantum number 3 and is the M shell. This can hold 2 electrons in the $3s$ orbital, 6 in the $3p$ and 10 in the $3d$. (Again the d orbitals are directed and have angular momentum.) The next higher energy level is the N shell which contains 2 electrons in the $4s$, 6 in the $4p$, 10 in the $4d$ and 14 in the $4f$ states. However, it happens that the lowest state in the N shell ($4s$) generally has lower energy than the $3d$ states in the M shell: electrons will therefore go into the $4s$ state before they start filling the $3d$ orbitals. The arrangement of shells and stable sub-groups are summarized in Table 1.

Table 1 *Shells and stable sub-groups*

Shell	Name	Number of electrons (maximum)	Number of electrons in stable sub-groups
1st	K	2	2
2nd	L	8	2, 6
3rd	M	18	2, 6, 10
4th	N	32	2, 6, 10, 14

As mentioned above if an atom contains a complete shell or sub-group it is relatively stable and may even be inert. In Table 2 we construct the first part of the periodic table based on the shell-model described above.

We see that the outermost electrons, especially those with unpaired spins, determine the chemical properties of the atoms. As a simple approximation we may say that the valency reflects the way in which the atom loses, gains or shares electrons in an attempt to form stable shells or sub-shells. When the atom already contains such a configuration (He, Ne, and Ar above) it is already stable and is chemically inert. This needs some qualification since in recent years it has been shown that the larger inert atoms such as xenon can, in fact, be ionized by the presence of a powerful electronegative ion such as the fluoride ion. As a result it is possible to form a stable compound of xenon hexafluoride.

9   Atoms, molecules and the forces between them

Table 2 *The beginning of the periodic table of elements*

Principal Quantum Numbers	1	2		3			4
Shell	K	L		M			N
Electronic states	1s	2s	2p	3s	3p	3d	4s 4p 4d 4f

At. No.	Atom	1s	2s	2p	3s	3p	3d	4s 4p 4d 4f
1	H	↑						
2	He	↑↓	inert					
3	Li	filled	↑					
4	Be	,,	↑↓					
5	B	,	filled	↑				
6	C	,,	,,	↑, ↑				
7	N	,,	,,	↑, ↑, ↑				
8	O	,,	,,	↑↓, ↑, ↑				
9	F	,,	,,	↑↓, ↑↓, ↑				
10	Ne	,,	,,	↑↓, ↑↓, ↑↓	inert			
11	Na	,,	,,	filled	↑			
12	Mg	,,	,,	,,	↑↓			
13	Al	,,	,,	,,	filled	↑		
18	Ar	,,	,,	,,	,,	↑↓, ↑↓, ↑↓	inert	
19	K	,,	,,	,,	,,	filled	empty	↑
20	Ca	,,	,,	,,	,,	,,	empty	↑↓
21	Sc	,,	,,	,,	,	,,	↑	↑↓
29	Cu	,,	,,	,,	,,	,,	filled	↑
Maximum occupancy number		2	2	6	2	6	10	2.6.10.14
Total		2	8		18			32

2·1·6 *Size of atoms*

The simplest way of forming some quantitative idea of the size of atoms is to measure the distance between atoms in the solid state. The result, of course, depends on the crystal structure. Some typical results are given in Tables 3 and 4.

Table 3 *Atomic diameters for vertical columns in periodic table*

| Atom | Z | Electron shells | | | | | | d Å. |
		K	L	M	N	O	P	
H	1	1						—
Li	3	2	1					3
Na	11	2	8	1				3·7
K	19	2	8	8	1			4·6
Rb	37	2	8	18	8			4·9
Cs	55	2	8	18	18	8	1	5·2

The outermost configuration is unchanged. As we fit in additional shells or subshells there is an increase in d. (The inner orbitals contract because of the increase in nuclear charge.)

Table 4 *Atomic diameters for horizontal columns in periodic table*

| Atom | Z | Electron shells | | | | d Å. |
		K	L	M	N	
Ca	20	2	8	8; 0	2	4
Sc	21	,,	,,	,, 1	2	3·5
Ti	22	,,	,,	,, 2	2	3·0
V	23	,,	,,	,, 3	2	2·6
Cr	24	,,	,,	,, 5	1	2·6
Mn	25	,	,,	,, 5	2	2·6
Fe	26	,,	,,	,, 6	2	2·5
Co	27	,,	,,	, 7	2	2·5
Ni	28	,,	,,	,, 8	2	2·5
Cu	29	,,	,,	,, 10	1	2·6

Note that Ir (Z = 77) has smaller diameter than Al (Z = 13). d (Ir) = 2·7 Å., d (Al) = 2·8 Å.

Here the main process is one of filling up an unfilled shell whilst the outer configuration scarcely changes. The increasing nuclear charge causes shrinkage in the atomic diameter.

11   Atoms, molecules and the forces between them

2·1·7 Size of ions

Some results are given in Table 5 for monovalent atoms which either lose an electron to become a positive ion, or gain an electron to become a negative ion. It is seen that the effect of removing one electron from the atom is to reduce d by 1·8 to 2 Å.; the effect of adding an electron is to increase d by about the same amount.

Table 5 *Effect of ionization on size*
(size obtained from spacing in appropriate compounds)

Metals				*Non-metals*		
	Diameter d Å.				*Diameter d* Å.	
Element	*Atom*	*Ion*		*Element*	*Atom*	*Ion*
Li	3	1·36		F	1·3	2·7
Na	3·7	1·96		Cl	2	3·6
K	4·6	2·96		Br	2·3	3·9
Rb	4·9	2·96		I	2·7	4·4
Cs	5·2	3·30				

One obtains a better idea of what is involved by comparing Na^+, Ne and F^-. These have the same final electron configuration, i.e. are isoelectronic but the nuclear charge is 11, 10, 9 respectively. Consequently one would expect the diameter of Na^+ to be smaller than Ne and that of F^- to be larger. This is indeed the case, the values being 2·0, 2·5 and 2·7 Å. for Na^+, Ne and F^- respectively.

2·1·8 Density of solid elements

An increase in atomic number usually implies an increase in atomic weight. There is also some increase or decrease in atomic diameter (see Tables 3 and 4) but it is not as large a change as the change in atomic weight. Consequently although there is generally an increase in density of the solid element as the atomic number increases it is not necessarily a monotonic relation.

It is interesting to note that if we compare the atomic weight of hydrogen with that of uranium (1 : 238) it is not very different from the ratio of their densities (0·07 : 18·4 g. cm.$^{-3}$).

2·2 The forces between atoms and molecules

2·2·1 *Forces due to the ionic bond*

The simplest forces between atoms are those which arise as a result of electron transfer. A simple example is that of, say, sodium fluoride. The sodium atom has a nuclear charge of $+11$, with 2 electrons in the K shell, 8 in the L shell and 1 in the M shell. The fluorine atom has a nuclear charge of 9 with 2 electrons in the K shell, and 7 in the L shell. The outermost electron in the sodium atom may transfer readily to the fluorine atom; both atoms then have a complete shell but the sodium now has a net charge of $+1$ and the fluorine a net charge of -1. These ions therefore attract one another by direct coulombic interaction. The force between them is strong – it varies as x^{-2}, where x is the distance between the ions, and it acts in the direction of the line joining the ions. Furthermore it is unsaturated – one positive ion can attract several negative ions around it and the force exerted by the positive ion on each negative ion is not affected by the presence of other negative ions. Of course the negative ions will also repel one another.

2·2·2 *Forces due to the covalent bond*

These are sometimes referred to as valency forces, for they are the type of force which accounts for the binding together of atoms in those molecules where the primary attraction is not ionic. For example, the force binding two hydrogen atoms together to form a hydrogen molecule is of this nature. Covalent bonding always involves the sharing of electrons. One way of describing this is to say that in the hydrogen molecule the single electron of each atom interacts with both nuclei, so that each atom regards itself as 'possessing' two electrons which provide a complete, stable 'molecular' K shell. However the binding cannot be described in classical terms. It arises from the fact that the electrons e_1 and e_2 are indistinguishable, so that if e_1 and e_2 are interchanged this does not involve a new configuration, but leads to an 'exchange energy' which provides the force binding the atoms

Figure 3. Electron concentration in a pair of hydrogen atoms; (*a*) unstable, the molecule dissociates into two atoms, (*b*) stable, the electron concentration between the nuclei serves to bond the nuclei together as the H_2 molecule.

together. Descriptively we may say that there is an appreciable concentration of electrons (negative charge) between the two nuclei (positive charge), and this binds the atoms together. This is indicated schematically in figure 3.

Covalent forces are strong, they fall off rapidly with separation; they act in specified directions (valency bonds) and they are saturated.

2·2·3 *Van der Waals or dispersion forces*

Van der Waals forces exist between all atoms and molecules, whatever other forces may also be involved. We may describe the origin of these forces by considering, for example, the interaction between two noble gas atoms, say two helium atoms. The atoms have complete K shells; they show no covalent bonding or ionic bonding; also since the electron shells are symmetrical about the nucleus the atom has no dipole. The latter statement is, however, only true as a time average. At any instant there may be some asymmetry in the distribution of the electrons around the nucleus. The helium atom at this instant, therefore, behaves as a dipole of moment μ (see

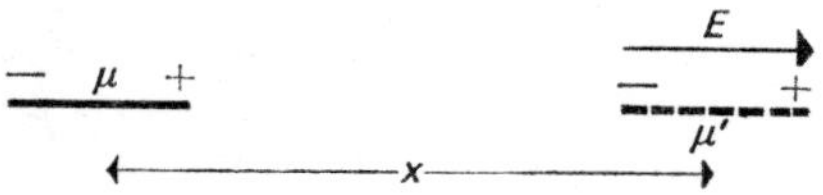

Figure 4. A dipole μ produces an electric field at a distance x along its axis of amount $E = \dfrac{2\mu}{x^3}$. This field can polarize another molecule such that mutual attraction occurs.

figure 4). At a distance x along the axis of the dipole the electrostatic field E is proportional to

$$\frac{\mu}{x^3} \quad \text{or} \quad E = \frac{k\mu}{x^3}. \tag{2.1}$$

If there is another atom at this point it will be polarized by the field and will acquire a dipole μ',

$$\mu' = \alpha E, \tag{2.2}$$

where α is the atomic polarizability of the atom.

Now a dipole μ' in a field E has a potential energy V given by

$$V = -\mu'E$$

$$= -\alpha E^2 = \frac{-\alpha k^2 \mu^2}{x^6}. \tag{2.3}$$

14   Gases, liquids and solids

This implies a force between the atoms of magnitude

$$F = \frac{-\partial V}{\partial x} = \frac{\text{constant}}{x^7}.$$ (2.4)

This derivation is rather crude since, of course, the second atom will also react on the first. The main point of this simplified treatment is to show that whatever the direction of the instantaneous electron fluctuation around one atom, it will always produce an attractive force on a neighbouring atom. The magnitude of the attractive force is also dependent on the frequency of these fluctuations. Indeed F. London (1930) showed that the potential energy between two identical atoms is given by

$$\text{P.E.} = -\frac{3}{4} \frac{\alpha^2}{x^6} h\nu,$$ (2.5)

where α is the polarizability, h is Planck's constant, and ν is the effective frequency of these fluctuations. As we shall see in Chapter 9 this frequency is the same as that which accounts for the dependence of refractive index of an assembly of atoms on the frequency of the incident light, i.e. it is the frequency involved in optical dispersion. For this reason the van der Waals forces are sometimes referred to in the older literature as dispersion forces.

Van der Waals forces are much weaker than ionic and covalent forces. The forces are central and like ionic forces they are unsaturated; also, to a first approximation, they are additive (see below).

2·2·4 *Retardation effects in relation to van der Waals forces*

The derivation of the London relation is valid only if the atoms are less than a few hundred angstroms apart; with larger distances the electromagnetic field from one atom takes an appreciable time to reach the next atom. For example, if the separation is 3000 Å. (3×10^{-5} cm.), the time taken is $\frac{3 \times 10^{-5}}{3 \times 10^{10}} = 10^{-15}$ sec. This is comparable with the period associated with the electron fluctuations; consequently there is a phase lag in the interaction. The theoretical treatment is very difficult but leads to the result that the force falls off as $1/x^8$ instead of $1/x^7$. This 'retardation' effect becomes important in studying the attraction between large solid bodies which are brought to within a small distance of one another. Since the van der Waals forces are roughly additive they may be integrated over the infinite half spaces. Allowing for retardation it turns out that for parallel surfaces, of separation h, the integrated force becomes proportional to $1/h^4$. The forces themselves, except when h is less than say 20 Å., are very small. For a separation of 5000 Å. the force is of the order of 10^{-6} g-wt cm.$^{-2}$ of surface. It is for this reason that one can safely say that the range of action of surface forces is small; in theory they extend to infinity, but their magnitude is appreciable only for a few tens of angstroms from the surface.

15 Atoms, molecules and the forces between them

The consequences of this may at once be seen. Surfaces will stick together strongly only if they can come into close contact (within a few angstroms) of one another. Again gases will generally be adsorbed strongly for the first monolayer but the attraction falls off so rapidly with distance that second and third layers are far less likely. Roughly speaking the formation of adsorbed layers depends on whether the potential energy drops by a greater amount than the mean thermal energy. For a molecule, the latter is of the order of kT, where k is the Boltzmann constant and T the absolute temperature (see Chapter 4). This is not the whole story since, of course, the latent heat of condensation is also a factor encouraging further adsorption, whilst the entropy decrease associated with an 'ordered' adsorbed film (see next chapter) tends to oppose adsorption. The net result is that polymolecular adsorption is unlikely unless the gas or vapour is very near saturation.

With large particles such as those in colloidal systems or Brownian suspensions the position is different. The thermal energy of the individual particle is still given by kT but the van der Waals energy must be integrated for the millions of atoms in the particles. In such systems the van der Waals forces can in fact play a basic part in determining whether particles will stick together or not, but of course a very small amount of electrostatic charge is likely to swamp the van der Waals forces even for colloidal systems.

2·2·5   *Repulsion forces*

If atoms were subjected only to attractive forces all atoms would coalesce. Thus it is clear that for very short distances of separation some type of repulsive force must operate. In ionic systems the attractive force arises from the electrostatic charge and the repulsive force from the electron shells, which act as a sort of tough elastic sphere resisting further compression. The repulsion may be described as arising from two effects. First, the penetration of one electron shell by the other, which means that the nuclear charges are no longer completely screened and therefore tend to repel one another. The other effect arises from the Pauli exclusion principle which states that two electrons of the same energy cannot occupy the same element of space. For them to be in the same space (overlapping) the energy of one must be increased – this is equivalent to a force of repulsion.

With covalent and van der Waals forces the repulsion and attraction are all part of a single mechanism and it is not correct to consider them as arising from separate mechanisms; however, for the purpose of simple calculation it is very convenient to do so. We therefore describe the long-range attraction by a term of the form

$$F = \frac{A}{x^m},$$
(2.6)

and the repulsion by a relation of the form

$$F = \frac{B}{x^n},$$ (2.7)

where B is a constant and n has a value of the order 8 to 10. In fact the power law relation is not very good and there are theoretical reasons for preferring a relation for the repulsive force of the form $Be^{-x/s}$ where B and s are suitable constants. In the following calculation we shall for simplicity retain the power law for both attraction and repulsion forces.

The resultant force is then

$$F = \frac{A}{x^m} - \frac{B}{x^n}$$ (2.8)

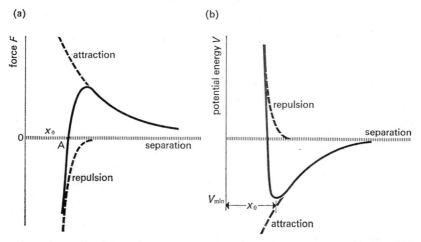

Figure 5. (a) Force and (b) potential energy curves for two atoms (or molecules) as a function of separation. There is a long-range attractive force and a short-range repulsive force which operates only when the molecules come close together. The equilibrium separation x_0 occurs when the net force is zero or when the potential energy is a minimum.

and this is plotted in figure 5. The equilibrium separation x_0 occurs when $F = 0$. This gives $B = Ax_0^{n-m}$.

Hence equation (2.8) becomes

$$F = A\left[\frac{1}{x^m} - \frac{x_0^{n-m}}{x^n}\right].$$ (2.9)

We may at once note that for small displacements from the equilibrium position the restoring force is

$$dF = A\left[\frac{n-m}{x_0^{m+1}}\right]dx.$$ (2.10)

17 Atoms, molecules and the forces between them

Since the force is proportional to the displacement the motion of the displaced particle will be simple harmonic. This conclusion is of course not restricted to power law relations. Any relation of the form $F = f(x)$ will lead to a similar conclusion.

2·2 6   *Potential energy*

The potential energy V involved in bringing one atom from infinity to a distance x from another atom is given by

$$V = \int (\text{external force})dx = - \int (\text{internal force})dx = \int Fdx, \quad (2.11)$$

since F acts in the opposite direction to dx.

From equation (2.9) this gives

$$V = A\left[-\underbrace{\frac{1}{m-1}\frac{1}{x^{m-1}}}_{\text{attractive term}} + \underbrace{\frac{1}{n-1}\frac{x_0^{n-m}}{x^{n-1}}}_{\text{repulsive term}} \right].$$

Note that the attractive term gives a negative potential energy, the repulsive term a positive potential energy (see figure 6).

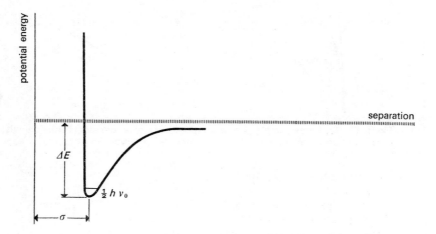

Figure 6. Stylized figure of potential energy curve of one molecule in relation to a single neighbour emphasizing the depth of the potential energy trough ΔE, the extremely steep repulsive part of the curve, and the effective diameter σ of the molecule. The zero-point energy, $\frac{1}{2}h\nu_0$, where ν_0 is the fundamental frequency of the pair, is very small compared with ΔE and will be neglected in the rest of this treatment.

At the equilibrium position where $x = x_0$ the potential energy has a minimum value

$$V_{min} = -A\frac{1}{x_0^{m-1}}\left[\frac{1}{m-1} - \frac{1}{n-1}\right].$$

The interatomic forces are summarized in Table 6.

Table 6 *Interatomic forces*

Type	Nature	A	m	Energy to separate 2 atoms or ions
Ionic	central unsaturated	very large	2	$Na^+ + F^-$; 10^{-11} erg
Covalent	directed saturated	very large	*	$H-H$; 7×10^{-12} erg
Van der Waals	central unsaturated	small to very small	7	$Ne:Ne$; 5×10^{-15} erg

* The attractive force between two hydrogen atoms in the hydrogen molecule is roughly proportional to $1/x^3$. This only applies for small displacements; it cannot be applied over large displacements nor can it be considered as being valid for covalent forces in general.

2·2·7   *Intermolecular forces*

Intermolecular forces are of the same nature as those occurring between atoms. For ionic crystals such as rock-salt one cannot distinguish individual molecules of NaCl, as the crystal is an array of positive (Na^+) and negative (Cl^-) ions held together by coulombic forces. In covalent solids such as diamond the individual carbon atoms are joined, by valency bonds, to form a strongly-linked crystal. In a solid such as paraffin wax, or solid hydrogen, the molecules are held together by weak van der Waals forces. However in a polymer these may be augmented by the attraction between polar groups or by chemical bonds. There is an additional type of bond which occurs only in the aggregate; this is the metallic bond. Individual metal atoms exist as such in the vapour phase, where a small number of diatomic molecules may occur, but there is very little tendency for the outermost valency electrons to join forces and produce polyatomic molecules in the vapour phase. There is however strong bonding in the condensed solid state. The atoms become ionized and exist as positive ions in a sea of free valency electrons; they are held together by the strong attraction between the ions and the electrons. In effect energy has to be provided to ionize the metal atoms, but this is more than compensated for by the binding energy between the ions and the electron sea.

19   Atoms, molecules and the forces between them

The simplest measure of the intermolecular forces is the heat of sublimation, which is the amount of energy needed to separate all the molecules from all their neighbours. This is highest for ionic and covalent solids, high for metals and least for van der Waals solids. We may make this more quantitative in the following way. If the heat of sublimation E_s is given in calories per gram-molecule, we multiply it by $4 \cdot 2 \times 10^7$ to convert to ergs and divide by 6×10^{23} to obtain the energy in ergs per single molecule. This conversion factor is 7×10^{-17}. The result is the energy E_s required to break bonds between one molecule and all its neighbours. If, therefore, the coordination number (number of nearest neighbours) is n and only nearest-neighbour interaction is involved, we may write for the binding energy E between one molecule and one neighbour

$$E_s = \tfrac{1}{2}nE.$$

The $\tfrac{1}{2}$ is introduced since otherwise every bond would be counted twice. We then have

$$E_s \times 7 \times 10^{-17} = \frac{n}{2}E \quad \text{or} \quad E = \frac{14 \times 10^{-17}}{n}E_s.$$

As an example we quote the simplest type of solid – solid neon, which forms a close-packed face-centred cubic (f.c.c.) structure of the van der Waals type. A sketch illustrating this structure is given in Chapter 8, figure 48. The latent heat of sublimation is found by experiment to be approximately 450 calories per gram-molecule. The number of nearest neighbours, n, is 12.

Hence

$$E \simeq 5 \times 10^{-15} \text{ erg.}$$

This agrees well with the theoretical interaction energy between one neon atom and its neighbour.

Another example, which is less direct, is worth quoting because it provides a check on our deduced value of E by a completely different method. In sodium fluoride the energy to convert 1 mole of NaF to Na^+ and F^- free ions is 213,000 cal. This is approximately 15×10^{-13} erg per molecule of NaF or $7 \cdot 5 \times 10^{-13}$ erg per ion of Na^+ or F^-. In the crystal lattice each ion is surrounded by 6 ions of opposite sign (i.e. $n = 6$). The energy per unit bond is therefore

$$E = \frac{7 \cdot 5 \times 10^{-12}}{3} = 2 \cdot 5 \times 10^{-12} \text{ erg.}$$

We may now consider the ionic interaction between a Na^+ and a F^- as they are separated from their equilibrium position in the lattice (separation $x_0 = 2 \cdot 31$ Å.) to infinity. Ignoring the repulsion term, we have

$$E = -\int_{x_0}^{\infty} \frac{e^2}{x^2}dx = -\frac{e^2}{x_0} = -10 \times 10^{-12} \text{ erg.}$$

This assumes only nearest-neighbour interaction. There is, in fact, a very large interaction between the next nearest neighbours. For example, the potential energy due to the attraction between one positive ion and the negative ions surrounding it, is greatly reduced by the coulombic repulsion between the positive ion and the nearest positive ions. These are only a little further away than the nearest negative charges and as coulombic potentials fall off only as $1/x$, the resultant potential energy is greatly reduced. Detailed calculations such as those indicated in Chapter 10 show that the energy is reduced from $-e^2/x_0$ to $0 \cdot 2905$ times this value. This gives a value for E of about $2 \cdot 9 \times 10^{-12}$ erg which agrees well with the value calculated from the heat of sublimation.

The main features of intermolecular forces are summarized in Table 7.

Table 7 *Intermolecular forces*

Type	Nature	Example	Heat to sublime solid to vapour. erg per bond $\times 10^{14}$	
Ionic	identical with interatomic forces	NaF crystal	break up into vapour of NaF molecules	300
Covalent	giant molecule held by covalent forces	diamond	break up into C atoms	300
Van der Waals	as for atoms	solid methane CH_4	break up into CH_4 gas	15
Metal	metal ions in sea of free valency electrons	Cu	vaporize to metal atoms	500

2·2·8 *Comparison of electrical and gravitational attraction*

Consider the interaction between, say, two helium atoms where only the weakest electrical attraction exists as a result of the van der Waals mechanism. If the atoms are packed together as close as possible (say in liquid state) the separation x_0 is about 2 Å. The potential energy V_w due to van der Waals for this separation turns out to be about -10^{-15} erg. Now consider the potential energy V_g due to gravity.

$$V_g = \int_{x_0}^{\infty} G \frac{m_1 m_2}{x^2} dx = -G \frac{m_1^2}{x_0} = -\frac{6 \cdot 7 \times 10^{-8}(0 \cdot 7 \times 10^{-23})^2}{2 \times 10^{-8}},$$

21 Atoms, molecules and the forces between them

c

i.e. $V_g = -10^{-46}$ erg

whereas $V_w = -10^{-15}$ erg.

We see that even when only van der Waals forces are involved the gravitational energy is trivial compared with the electrical.

2·2·9 *The simplified potential energy curve*

In this section we shall treat molecules as though they are hard elastic balls of diameter σ which attract each other according to some suitable law of force, and then experience a very powerful repulsion force for separations less than σ.

The resulting potential energy curve is then as shown in figure 6 and the two most important factors are σ and the depth $\Delta\varepsilon$ of the potential energy trough.

Some typical values are as follows in Table 8.

Table 8

Molecule	σ Å.	$\Delta\varepsilon$ erg (approx.)
He	2·2	1×10^{-15}
H_2	2·7	4×10^{-15}
Ar	3·2	5×10^{-15}
N_2	3·7	13×10^{-15}
CO_2	4·5	40×10^{-15}

In later chapters we shall be able to explain, in terms of such a potential energy curve, a large number of phenomena. These include critical temperature of a gas, surface tension and viscosity of a liquid, heat of sublimation of a solid, elastic constants and the thermal expansion of a solid.

Postscript on electronic energy levels

The fact that the simple kinetic theory of gases is valid sheds light on the electronic energy levels in the atom. Consider for simplicity a monatomic gas. The whole of the gas behaviour is consistent with the view that the only energies involved are the kinetic energy ($\frac{3}{2}tK$) of the atom as a whole (see Chapter 5). Collisions do not share the kinetic energy with say the electronic energy of electrons within the atom. The reason (see N. F. Mott, *Contempory Physics*, vol. 5, 1964, p. 401) is that the electronic energies are quantized and the gap between the various energy levels is enormous compared with the ordinary kinetic energy due to thermal motion (5 eV compared with 0·02 eV at room temperature). For the same reason the electronic energies play no part in the specific heat. Ordinary temperatures are far too small to promote the electrons from one energy level to another. This has a direct bearing on the Langevin treatment of diamagnetism (see pp. 258, 269).

Chapter 3
Temperature, heat and the laws of thermodynamics

In this chapter we consider briefly some of the basic concepts of heat and temperature. The treatment is simple and restricted to those aspects which will be of use in later parts of the book.

3·1 Temperature

3·1·1 *The concept of temperature*

The concept of temperature originally arose from a sensory feeling of hot or cold. Probably all physical concepts arise in a similar way; the next step is to turn the idea into something more general and, in particular, more objective. One such approach is as follows. It is found that a given mass of gas is completely specified by its volume V and its pressure P. Suppose we start with 1 kg. of a gas and subject it to any changes we wish whether by compressing it, cooling it, allowing it to expand heating it; we then take another 1 kg. specimen of the same gas and carry out a completely different series of operations. If we end up with the same values of V and P for the two samples, the final states will be found identical in every way – colour, warmth, viscosity, and every objective and subjective test to which we can subject the sample.

If we end up with different P, V values there is something different about the samples and if they are placed in contact changes take place until they reach identical P, V values (see section 3·1·3 below).

The attribute which produces this difference is the temperature and the changes brought about when the specimens are placed in contact result in their eventually reaching the same temperature. For gases and fluids there is a unique function of (P, V) which determines their temperature θ if the mass is specified: $f(P, V) = \theta$. For solids one needs more parameters.

3·1·2 *Temperature scales*

There are, of course, sophisticated means of deriving an absolute thermodynamic scale of temperature (see section 3·3·6 below). In practice one can use the P–V properties of a gas. One may also use less direct standards such as the expansion of a metal rod, the expansion of a liquid column (a conventional thermometer), or the e.m.f. produced by a thermocouple.

3·1·3 Zeroth law of thermodynamics

In the development of thermodynamics the first two laws were developed and so named, before it was realized that a preliminary law was required to satisfy a rigorous formulation of thermodynamic laws. This was therefore called the zeroth law. (It has nothing to do with Professor Zero.) The law simply states that if two bodies A and B are individually in thermal equilibrium with a body C, then A and B are in thermal equilibrium with one another.

If C is a thermometer this is clearly a matter of common experience.

3·2 Heat

3·2·1 The concept of heat

Suppose we have a body of mass M_1 at temperature θ_1, which is placed in thermal contact with another body of mass M_2 and temperature θ_2, where θ_2 is greater than θ_1. Equilibrium will be reached at a common temperature θ, $\theta_2 > \theta > \theta_1$. We say that heat has flowed from θ_2 to θ_1.

Without specifying what heat is, we can define its units. The heat required to raise the temperature of 1 gram of water at 15°C. by 1°C. is the calorie. If we assume that in any experiment involving the flow of heat from one body to another heat is conserved we may define the specific heat of a given substance in terms of the specific heat of, say, water. The specific heat s is defined as the number of calories required to raise the temperature of one gram of the substance by 1°C.

We can now carry out experiments with different masses of different materials at various temperatures using only a centigrade scale thermometer. Common experience then shows that all the results are consistent with our original assumption that heat is always exactly conserved. We have to qualify this by adding that this is true only if any volume changes which occur do a negligible amount of work.

So far we have said nothing about the nature of heat, but this does not affect the validity of the calorimetric conservation principle which we have just described. The next step is to consider briefly J. P. Joule's classical experiments which show that heat is, in fact, a form of energy.

3·2·2 Joule's experiments

J. P. Joule (1818–1889) began his famous series of experiments on the relation between heat and energy in 1840 and summarized his major conclusions in a massive mémoire published in the *Philosophical Transactions* in 1850. His first experiments (1843) involved the combined effects of mechanical and electrical energy, while his second experiments (1845) were concerned with the heat liberated when a gas is compressed. His more famous experiment in which he directly compared the mechanical work expended in stirring water with the heat evolved did not appear until later in 1845, and a greatly improved version was described by him in 1878. The principle was to stir water in a copper vessel with paddles and to calculate the mechanical work done W. Using a very sensitive thermometer he observed the temperature rise Δt of the water, paddles and vessel, and estimated heat losses by conduction, radiation and (if relevant in the mechanical arrangements of the driving system) the losses due to friction in the bearings. From subsidiary calorimetric experiments he was able to determine the specific heats of the vessel and the paddles in terms of the unit heat capacity of water. In this way he could calculate the total amount of heat H in a purely calorimetric experiment required to produce the same temperature rise as in the paddle-stirring experiment, viz:

$$H = \Delta t \text{ [Mass of water} + m_1 s_1 \text{ (vessel)} + m_2 s_2 \text{ (paddles)]}$$
$$+ \text{ losses.} \tag{3.1}$$

He then found that H was closely proportional to the mechanical work W. We may summarize some of his other experiments.

(a) He repeated the stirring experiments with mercury as the liquid.
(b) He rubbed iron rings together against their interfacial friction.
(c) He passed an electric current through a resistance wire immersed in a liquid.
(d) He compressed air in an insulated cylinder.

In all cases he found that the observed temperature rises could be achieved simply by adding heat of an amount proportional to the mechanical or electrical work done. Further, the factor of proportionality was a constant

$$W = JH \tag{3.2}$$

where J is Joule's constant or the mechanical equivalent of heat and has the value $J = 4\cdot187 \times 10^7$ erg calorie^{-1}.

Joule's constant is of immense importance in all mechanical–thermal operations, but in effect this reduces to translating specific heat data into energy data. It is clearly not a fundamental unit of physics. Indeed if calorimetric studies (involving pounds or grams of water and degrees centigrade or fahrenheit) had not been so attractive as a convenient means of describing thermal energy, it is possible that our basic unit would have been the Joule (10^7 erg); we should then have had a calorimetric heat-unit defined as the heat required to raise the temperature of one gram of water by $0\cdot238°C$.

3·3 The laws of thermodynamics

3·3·1 *Internal energy and the first law of thermodynamics*

If we carry out an experiment in a thermally isolated calorimeter, such that no heat can flow into or out of the system, we can change the state of the

25 Temperature, heat and the laws of thermodynamics

system by performing mechanical or electrical work on it. Joule's experiments show that for the same amount of work we shall always arrive at the same final state. We may now define the difference in internal energy U of the system as the work done on the system in an adiabatic calorimeter.

Since we can change from state 1 to state 2 by an infinite number of combinations of heat and work, if we wish to define internal energy as a function of state only, we must include both work and heat in our definition of energy. The following example illustrates this in a more concrete way.

Suppose we take a mass of water and perform mechanical and/or electrical work on it; provided we do the same amount of work W, we produce the same change in temperature. Alternatively we can apply a quantity of heat $H = W/J$ and produce the same change. If we start with water at 0°C. and end with water at 100°C., then the water ends with the same energy state whatever the path adopted. For example, we could add heat to raise the temperature to 30°C. and then stir paddles to supply the remaining 70°C. rise. Clearly an infinite number of paths are available. Yet the change in internal energy $U_{100} - U_0$ (ignoring any work done by the liquid on expanding) must be determined only by the initial and final temperature, not by the path. If this were not true we should be able to devise heating and cooling cycles whereby we could raise the temperature to 100°C. by one path and cool it back to 0°C. by another, but still obtain energy from the system. This is contrary to Joule's experiments.

We are left with the conclusion that a change in U is a function only of the initial and final states. If we do external work ΔW and add heat ΔQ we have

$$\Delta U = \Delta W + \Delta Q, \tag{3.3}$$

where ΔU is uniquely defined, but ΔW and ΔQ can have any values provided their sum is equal to ΔU.

For a specified change in state

$$W_A + Q_A \text{ over path A} = W_B + Q_B \text{ over path B.} \tag{3.4}$$

Thus work plus heat are conserved. We may regard this as an extension of the conservation of energy which is well established in *mechanics* and conclude that:

Energy is conserved if we include heat as one of the forms of energy.

This is one way of formulating the first law of thermodynamics.

In the simple example given above we have considered U to consist of heat and mechanical or electrical energy. A little consideration shows that the internal energy can embrace all forms of energy, thermal, mechanical, electrical, gravitational, electromagnetic, etc. It may again be shown by similar arguments that, however many types of energy are involved, any change in the internal energy does not depend on the path; it is a function solely of the initial and final states.

3·3·2 *Internal energy of fluids*

For a fluid, U is found to be a unique function of only 2 variables; either (P, V), (P, T) or (V, T), where P is the pressure, V the volume and T the temperature. Since U is independent of path, changes in it can always be expressed in terms of an exact differential. This also means that, in principle, we can integrate dU between initial and final states and always end up with the same answer. Thus we can always write

for $U(T, V)$

$$dU = \left(\frac{\partial U}{\partial T}\right)_V dT + \left(\frac{\partial U}{\partial V}\right)_T dV, \qquad (3.5a)$$

for $U(P, V)$

$$dU = \left(\frac{\partial U}{\partial P}\right)_V dP + \left(\frac{\partial U}{\partial V}\right)_P dV. \qquad (3.5b)$$

In our example of raising the temperature of water from 0°C. to 100°C. we concentrated on the variation of U with temperature; $\left(\dfrac{\partial U}{\partial T}\right)_V$ in equation (3.5a). That is why we made the proviso that for this to be strictly true any changes in volume would have to be ignored, i.e., $\left(\dfrac{\partial U}{\partial V}\right)_T = 0$.

3·3·3 *Reversible changes in a gas*

If we allow a gas to expand, it does work against its surroundings. The work done depends on the pressure of the gas, the volume change and the pressure outside the gas against which it does work.

Consider a cylinder of cross-sectional area A closed with a weightless piston. The work done during an expansion can be defined uniquely only if

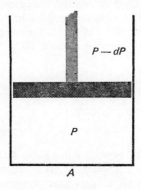

Figure 7. Reversible expansion of a gas.

(*a*) the piston is frictionless, (*b*) the expansion is slow and always in quasi-equilibrium, i.e. the pressure outside the piston is only infinitesimally smaller than P so that, by a minute increase in the external pressure, we would be able to *compress* the gas. Under these conditions the gas pushes with a force PA against the surroundings and if the piston moves a distance dx the work done is $PAdx = PdV$, where dV is the volume increase. For such a *frictionless, reversible* expansion PdV is uniquely defined for the increment dV.

Consider now what happens if we increase the internal energy of a gas by allowing it to do work w, and also add heat of amount q. We have

$$\underset{\text{unique}}{dU} = \underset{\text{not unique}}{-w} + \underset{\text{not unique}}{q}. \tag{3.6}$$

Here w is the work done *by* the gas and is therefore counted negative. For reasons, which we described before, dU is determined uniquely by the initial and final states. This is not generally true of $-w$ and q separately and for this reason they cannot be expressed as exact differentials.

If we now specify that the work w is a reversible frictionless expansion we have

$$\underset{\text{unique}}{dU} = \underset{\text{unique}}{-PdV} +q. \tag{3.7}$$

Consequently q must also be uniquely defined over this increment of change so that it can be written as dQ.

Hence

$$dQ = dU + PdV. \tag{3.8}$$

Each of the quantities in equation (3.8) is now in the form of a differential, so if the relevant paths are known, they can all be integrated exactly. We see now why frictionless reversible expansions or compressions are so important in thermodynamics. In an electrochemical system the term equivalent to PdV would be Edq where E is the potential difference and dq the amount of charge passing. Here again for an equation like (3.8) to be valid the current flow would have to take place against an opposing potential infinitesimally less than E. In a magnetization process, where a magnetic element of moment M is exposed to a magnetic field dH, the equivalent term would be MdH.

3·3·4 *Enthalpy*

We see from equation (3.8) that the internal energy may be written

$$dU = dQ - PdV. \tag{3.8a}$$

Unfortunately we cannot integrate this, unless we know the equation of state and the path followed by the expansion cycle. For example an adiabatic expansion would give a value of $\int PdV$ which differs from that occurring in an isothermal expansion. If, however, we specify that the expansion is maintained at constant pressure we have

$$U_2 - U_1 = Q_2 - Q_1 - P(V_2 - V_1). \tag{3.9}$$

The heat absorbed at constant pressure is then given by

$$Q_2 - Q_1 = (U_2 + PV_2) - (U_1 + PV_1). \tag{3.10}$$

We now introduce a new thermodynamic quantity called enthalpy which is defined as

$$H = U + PV. \tag{3.11}$$

This is also a function solely of initial and final states. Thus for a small change

$$dH = dU + PdV + VdP = dQ + VdP. \tag{3.12}$$

Hence

$$\left(\frac{\partial H}{\partial T}\right)_P = \left(\frac{\partial Q}{\partial T}\right)_P. \tag{3.13}$$

As we shall see later enthalpy is conserved in the steady flow of a gas through a throttle (Joule–Kelvin expansion).

3·3·5 *Specific heats*

These may be defined as the amount of heat required to raise the temperature of a specified amount of material (one gram, one gram-molecule) by a unit degree of temperature:

$$C = \frac{dQ}{dT}. \tag{3.14}$$

In general we can specify two different conditions for the measurement of C.

Constant volume

$$C_V = \left(\frac{\partial Q}{\partial T}\right)_V = \left(\frac{\partial U}{\partial T}\right)_V \quad \text{from equation (3.8a).}$$

Constant pressure

$$C_P = \left(\frac{\partial Q}{\partial T}\right)_P = \left(\frac{\partial H}{\partial T}\right)_P. \tag{3.15}$$

For gases, as we shall see, there is an appreciable difference between C_P and C_V. For liquids and solids the difference is very much smaller. Some typical values of C for metals are given in Table 9.

Table 9 *Specific heat of some metals*

Metal	Specific heat C cal. g.$^{-1}$ °C.$^{-1}$	Atomic weight	Product
Al	0·22	27	6
Cu	0·09	63·5	5·7
As	0·054	108	5·8
Pb	0·03	207	6·2

The last column shows that the product of the specific heat and the atomic weight (this is in effect the specific heat per gram atom) has a value of about 6 cal. g-atom^{-1} °C.$^{-1}$. This conclusion is known as the law of Dulong and Petit and we shall discuss its basis and validity in a later chapter.

3·3·6   *The second law of thermodynamics*

It is not the purpose of this book to provide a complete account of thermo- dynamics (readers will find a fascinating and stimulating account in A. B Pippard (1957), *Elements of Classical Thermodynamics*, Cambridge University Press). Indeed we shall be able to deal with practically all our needs using only the first law. There is, however, some point in emphasizing the essential difference between heat and other forms of energy. When a force does work, the work done is the product of force and distance moved, the force always being measured in the direction of the line along which it is acting. When an electric current does work the current flows in the same direction as the resultant e.m.f. Mechanical work and electrical work are in principle absolutely convertible into one another. A dynamo in principle produces an amount of electrical power exactly equal to the work done in driving it (apart from friction and electromagnetic losses). If this is converted into heat there is, again, an exact equivalence as Joule showed. If, however, we wish to convert heat into work this is *not* true, because thermal energy always has associated with it random movement, vibration, thermal agitation, so there is always part of the thermal energy which is not free for conversion into mechanical or electrical energy.

N. L. S. Carnot (1824) showed that if one had a heat engine working between a source at a higher temperature and a sink at a lower temperature, the engine would achieve its greatest efficiency if all the parts of the cycle were carried out reversibly. Under these conditions the efficiency was independent of the working material. By efficiency we mean the ratio of the useful work done to the thermal energy absorbed from the higher temperature source. For reversible cycles if the heat taken in is Q_1 and the heat rejected to the sink is Q_2, the efficiency η is

$$\eta = \frac{Q_1 - Q_2}{Q_1} .\tag{3.16}$$

Since this is independent of the working material it depends only on the temperature of the two sources. This provides a means of establishing a thermodynamic scale of temperature, which is achieved by defining the ratio of heat absorbed at the higher temperature to that rejected at the lower temperature as equal to the ratio of the higher to the lower temperature, i.e., $\frac{Q_1}{Q_2} = \frac{T_1}{T_2}$. For such an absolute scale of temperature

$$\frac{Q_1 - Q_2}{Q_1} = \frac{T_1 - T_2}{T_1} .\tag{3.17}$$

We see that fundamentally the efficiency can never be unity unless T_2, the temperature of the sink, is zero. Under these conditions the heat, as it were, is acting completely in the direction of doing useful work.

For the reversible cycles we see that

$$\frac{Q_1}{T_1} = \frac{Q_2}{T_2} .\tag{3.18}$$

Hence travelling from the initial condition, through the cycle, and then back to the initial condition again

$$\frac{Q_1}{T_1} - \frac{Q_2}{T_2} = 0 .\tag{3.19}$$

We may now define dQ/T as the entropy change dS in a reversible process. The word entropy was invented by R. Clausius; it comes from the Greek word *trope* meaning transformation and he deliberately added the prefix *en* to convey a connection with energy. For a reversible cycle (the sign $\oint$ indicates that the integral is taken over the complete cycle)

$$\oint \frac{dQ}{T} = \oint dS = 0 .\tag{3.20}$$

This implies that dS is an exact differential; for any increment in a reversible process

$$dS = \frac{dQ}{T} \quad \text{or} \quad dQ = TdS .\tag{3.21}$$

One of the most interesting properties of entropy is the following. Even for a reversible cycle the integral of dQ depends on the details of the paths followed (see discussion under enthalpy). If, however, we divide the increment

31   Temperature, heat and the law of thermodynamics

of dQ by the temperature at which the heat dQ is absorbed (or rejected), the quantity dQ/T in any reversible path can be integrated between the initial and final state and is independent of the path followed. This is what is meant by saying that dS is an exact differential.

For completeness we quote two other thermodynamic functions which are independent of path, the Helmholtz free energy A for changes at constant volume, and the Gibbs free energy G for changes at constant pressure. They are defined as follows:

$$A = U - TS \tag{3.22}$$

$$G = H - TS. \tag{3.23}$$

We see that for an isothermal change

$$dA = dU - TdS. \tag{3.24}$$

The change dA in free energy A is equal to the change in internal energy U less the 'unavailable' energy TdS. This is the maximum work that the system can perform. For a non-reversible process the available work will be less than dA. At constant temperature and constant pressure the maximum work that the system can perform is dG; this corresponds to the change in enthalpy less the unavailable energy TdS.

It is not surprising, since the unavailability of thermal energy is associated with its randomness, that entropy should in some way be connected with the probability or distribution-randomness of the system under consideration. What is surprising is that this concept can be made quantitative. If the probability of finding a system in a given state is W_1 and in another state is W_2, the entropy change may be written

$$\Delta S = k \ln \frac{W_2}{W_1}, \tag{3.25}$$

where k is the Boltzmann constant. Of course the probability needs to be correctly defined and this forms the subject of statistical mechanics. The probability associated with the translational motion of gas molecules turns out to be proportional to the volume occupied by the gas; the larger the volume the greater the randomness. The entropy change in increasing the volume from V_1 to V_2 at constant temperature is

$$\Delta S = k \ln \left(\frac{V_2}{V_1} \right)^N,$$

where N is the number of molecules in the quantity of gas considered. We shall use this result in a later chapter of the book.

Chapter 4
Perfect gases – bulk properties and simple theory

4·1 Bulk properties

4·1·1 Summary of main bulk properties

As we shall see in the course of this and the next two chapters, all gases will approximate to perfect or ideal gases if they are sufficiently dilute and if the intermolecular forces are negligible compared with the thermal energy. This condition applies reasonably well to oxygen, nitrogen and hydrogen at normal temperatures and pressures. For this reason the earliest studies of the 'springiness' of air were concerned with the behaviour of a perfect gas.

The main bulk properties were established long ago. We summarize the main conclusions:

1. Boyle's law (*ca.* 1660). For a given mass of gas at a fixed temperature the product of pressure and volume is a constant.

$$PV = \text{constant.} \tag{4.1}$$

2. Charles' law (1787) or Gay-Lussac's law (1802). For a given mass of gas, if the pressure be kept constant the volume increases linearly with the temperature.

$$V = V_0(1 + \alpha t). \tag{4.2}$$

If the temperature is measured on the centigrade scale and V_0 is the volume at 0°C., it is found that α has the value $\alpha = \frac{1}{273}$. We may therefore write equation (4.2) as

$$V = V_0\left(1 + \frac{t}{273}\right) = \frac{V_0}{273}(273 + t)$$

$$= \frac{V_0}{273}T, \tag{4.3}$$

where T is a scale of temperature, which has -273°C as its zero point. This scale turns out to be the same as the thermodynamic scale.

3. Variation of pressure with temperature if the volume is kept constant.

$$P = P_0(1 + \beta t). \tag{4.4}$$

The constant β has the same value of $\frac{1}{273}$ so that we have

$$P = \frac{P_0}{273} T.$$ (4.5)

4. General gas equation. Rewriting equation (4.1) in the light of equations (4.3) and (4.5) we obtain a general equation of state

$$PV = RT,$$ (4.6)

where R depends only on the quantity of gas used.

5. Avogadro's law. For all perfect gases R has the same value if the same molecular quantity of gas is considered. For one gram-molecule

$$R \simeq 8 \times 10^7 \text{ erg } °K.^{-1}$$

$$\simeq 2 \text{ cal. } °K.^{-1}$$

6. Dalton's law. This states that, at a fixed temperature, the pressure of a mixture of gases is equal to the sum of the pressures which would be exerted by each gas separately, if the other constituents were not there.

4·1·2 *Reversible isothermal expansion*

We allow a gas to expand in a frictionless cylinder against a pressure infinitesimally less than the gas pressure. The temperature T is maintained constant. The gas does work

$$dW = PdV.$$ (4.7)

For an ideal gas $PV = RT$ so that P in equation (4.7) can be replaced by $\frac{RT}{V}$.

Then $$dW = \frac{RT}{V} dV.$$

If expansion occurs, from volume V_1 and pressure P_1 to volume V_2 and pressure P_2, the work done is

$$W = RT \ln \frac{V_2}{V_1} = RT \ln \frac{P_1}{P_2}.$$ (4.8)

This work is provided by the heat absorbed from the constant temperature source, which must be included in the system if T is to be kept constant.

4·1·3 *Fast adiabatic expansion into a vacuum*

Consider the experimental arrangement shown in figure 8. A volume V of gas is maintained in the left-hand flask at pressure P and temperature T.

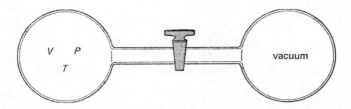

Figure 8. Fast adiabatic expansion of a gas into a vacuum. Joule found that for dry air there was no net temperature change.

The right-hand flask is evacuated. The whole system is thermally insulated so that heat cannot flow into or out of the flasks, then the tap is opened and the gas rushes into the empty flask. Some inequalities of temperature occur initially but a steady state is quickly reached. Joule found that for dry air the temperature in the steady state is indistinguishable from the original temperature. We write

$$dU = \Delta Q + \Delta W. \tag{4.9}$$

For an adiabatic process there is no heat flow, $\Delta Q = 0$, and for an expansion against zero pressure no work is done, i.e. $\Delta W = 0$.
　　Hence

$$dU = 0. \tag{4.10}$$

This means that U is unchanged and is independent of volume occupied (or of pressure exerted). This implies that there are negligible forces between the gas molecules; otherwise the internal energy would depend on how far we had separated the molecules, i.e. on the volume.
　　Analytically we may write

$$U = f(V, T) \quad \text{or} \quad (P, T)$$

$$dU = \left(\frac{\partial U}{\partial V}\right)_T dV + \left(\frac{\partial U}{\partial T}\right)_V dT.$$

Since there is no temperature rise, $dT = 0$; also from equation (4.10), $dU = 0$, so consequently

$$\left(\frac{\partial U}{\partial V}\right)_T = 0, \quad \text{and similarly} \quad \left(\frac{\partial U}{\partial P}\right)_T = 0. \tag{4.11}$$

Thus for a perfect gas U is a function of T only and neither P nor V is involved.

4·1·4　*Specific heat of a gram-molecule of a perfect gas*

Consider a small equilibrium change in the gas if we increase its temperature by dT. The basic equation is

$$dQ = dU + PdV$$

$$= \left(\frac{\partial U}{\partial V}\right)_T dV + \left(\frac{\partial U}{\partial T}\right)_V dT + PdV.$$

For a perfect gas the first term on the R.H.S. is zero. Thus we have

$$dQ = \left(\frac{\partial U}{\partial T}\right)_V dT + PdV. \tag{4.12}$$

Consider now two different conditions.

(a) *Constant volume*: $dV = 0$. We can rewrite equation (4.12) as

$$dQ = \left(\frac{\partial U}{\partial T}\right)_V dT.$$

Hence

$$\left(\frac{\partial Q}{\partial T}\right)_V = C_V = \left(\frac{\partial U}{\partial T}\right)_V. \tag{4.13}$$

(b) *Constant pressure*: The gas expands during heating. Equation (4.12) now becomes

$$dQ = C_V dT + PdV.$$

Hence

$$\left(\frac{\partial Q}{\partial T}\right)_P = C_V + P\left(\frac{\partial V}{\partial T}\right)_P. \tag{4.14}$$

For a perfect gas $PV = RT$ so that $PdV + VdP = RdT$. At constant pressure $dP = 0$, consequently

$$P\left(\frac{\partial V}{\partial T}\right)_P = R.$$

So equation (4.14) becomes

$$\left(\frac{\partial Q}{\partial T}\right)_P = C_P = C_V + R,$$

or $C_P - C_V = R.$ \tag{4.15}

4·1·5 *Reversible adiabatic expansion*

We start with our basic equation

$$dQ = dU + PdV,$$

specifying that, for adiabatic process, $dQ = 0$. Our starting equation then becomes

$$0 = C_V dT + PdV. \tag{4.16}$$

For a perfect gas, the gas equation always applies:

$$PdV + VdP = RdT.$$

Equation (4.16) becomes:

$$- C_V dT = PdV = RdT - VdP$$
$$= (C_P - C_V)dT - VdP.$$

Hence $\qquad C_P dT = VdP.$ $\qquad\qquad$ (4.17)

From (4.16)

$$C_V dT = - PdV. \qquad\qquad (4.18)$$

Taking the ratio of equations (4.17) and (4.18)

$$\frac{C_P}{C_V} = \gamma = \frac{-VdP}{PdV},$$

or $\qquad \gamma \dfrac{dV}{V} = - \dfrac{dP}{P}.$

Integrating from initial to final conditions

$$\gamma \ln \frac{V_2}{V_1} = - \ln \frac{P_2}{P_1}$$

or $\qquad \ln \left(\dfrac{V_2}{V_1} \right)^{\gamma} + \ln \left(\dfrac{P_2}{P_1} \right) = 0$

or $\qquad \ln \dfrac{P_2 V_2^{\gamma}}{P_1 V_1^{\gamma}} = 0.$ $\qquad\qquad$ (4.19)

Since the antilogarithm of 0 is 1 this gives

$$P_1 V_1^{\gamma} = P_2 V_2^{\gamma} = \text{constant.} \qquad\qquad (4.20)$$

If we plot the pressure against the volume we see (figure 9) that the adiabatic curve is steeper than the isothermal.

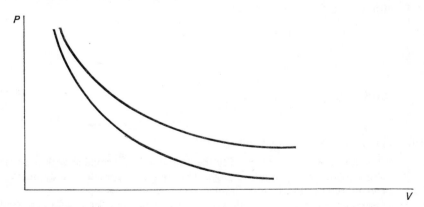

Figure 9. Pressure–volume relations for a perfect gas. The adiabatic (lower) curve is steeper than the isothermal (upper) curve.

37 Perfect gases – bulk properties and simple theory

4·1·6 Work done in reversible adiabatic expansion

The work done by the gas in expanding from P_1V_1 to P_2V_2 is

$$dW = PdV. \tag{4.21}$$

At any stage we can write

$$PV^\gamma = P_1V_1^\gamma$$

so that P in equation (4.21) can be replaced by $P_1V_1^\gamma V^{-\gamma}$. The equation becomes

$$W = \int_{V_1}^{V_2} P_1V_1^\gamma V^{-\gamma}dV.$$

Therefore

$$W = \frac{P_1V_1}{\gamma-1}\left[1-\left(\frac{V_1}{V_2}\right)^{\gamma-1}\right] = \frac{P_1V_1}{\gamma-1}\left[1-\left(\frac{P_2}{P_1}\right)^{\frac{\gamma-1}{\gamma}}\right]. \tag{4.22}$$

4·1·7 Cooling as a result of adiabatic expansion

The work done by the gas is done at the expense of the internal energy. Instead of calculating this *de novo* we merely combine the adiabatic expansion equation with the general gas equation:

$$P_1V_1^\gamma = P_2V_2^\gamma,$$

$$\frac{P_1V_1}{T_1} = \frac{P_2V_2}{T_2}. \tag{4.23}$$

Taking the ratio of the L.H.S. of the two equations and equating it to the ratio of the R.H.S. we obtain

$$\frac{P_1V_1^\gamma}{\dfrac{P_1V_1}{T_1}} = \frac{P_2V_2^\gamma}{\dfrac{P_2V_2}{T_2}}.$$

Hence

$$\frac{T_2}{T_1} = \left(\frac{V_1}{V_2}\right)^{\gamma-1} = \left(\frac{P_2}{P_1}\right)^{\frac{\gamma-1}{\gamma}}. \tag{4.24}$$

4·1·8 Entropy changes on expansion

When a gas expands isothermally the molecules are provided with a greater volume without any other change. This implies a greater randomness or probability and this, in turn, implies an increase in entropy.

For an isothermal expansion the entropy change for a gram-molecule of gas is easy to calculate, though the proof given here cannot be considered very rigorous. If the initial volume is V_1 and the final volume V_2 and there are N molecules, the probability of finding one molecule in the final volume is $\dfrac{V_2}{V_1}$ times greater than that of finding it in V_1. For the N molecules the relative probabilities of finding all the molecules in V_2 compared with V_1 is $\left(\dfrac{V_2}{V_1}\right)^N$. The change in entropy is then

$$\Delta S = k \ln \frac{\text{probability 2}}{\text{probability 1}} = k \ln \left(\frac{V_2}{V_1}\right)^N$$

$$= kN \left(\ln \frac{V_2}{V_1}\right) = R \ln \left(\frac{V_2}{V_1}\right), \tag{4.25}$$

if we are dealing with a mole of gas, i.e. if $N =$ Avogadro's number, N_0.

For the adiabatic expansion we may divide the process into two independent reversible parts. We first allow the gas to expand isothermally from V_1 to V_2, giving the entropy increase derived above. We then cool the gas (at constant volume V_2) until its temperature is reduced from T_1 to the final (adiabatic) temperature T_2. The decrease in entropy for such a reversible cooling is

$$\Delta S = \int \frac{dq}{T} = \int_{T_1}^{T_2} \frac{C_V dT}{T} = C_V \ln \frac{T_2}{T_1}. \tag{4.25a}$$

But as we saw in equation (4.24) $\dfrac{T_2}{T_1}$ for the adiabatic expansion is equal to $\left(\dfrac{V_1}{V_2}\right)^{\gamma-1}$. The entropy loss on cooling is therefore

$$C_V \ln \frac{T_2}{T_1} = C_V(\gamma-1)\left(\ln \frac{V_1}{V_2}\right) = C_V\left[\frac{C_P}{C_V}-1\right]\ln \frac{V_1}{V_2}$$

$$= (C_P - C_V)\ln \left(\frac{V_1}{V_2}\right)$$

$$= R \ln \frac{V_1}{V_2} = -R \ln \frac{V_2}{V_1}. \tag{4.25b}$$

If we add this to the entropy change associated with the isothermal expansion part of the process [equation (4.25)], we see that the total entropy change is zero. This is because the volume increase tends to increase the randomness whereas the temperature decrease tends to reduce it, the two effects exactly balancing out. This is, of course, what we should expect, since in the adiabatic process itself no heat is added or withdrawn from the gas, so $\Delta Q = 0$ and therefore $\Delta S = \int \dfrac{dQ}{T}$ must be zero.

39   Perfect gases – bulk properties and simple theory

The pressure of the atmosphere diminishes with height and in the next chapter we shall explain why. The atmospheric pressure drops by about $\frac{1}{30}$th part for every 1000 ft of ascent. If at ground level the pressure is P_0, the pressure 1000 ft up is about $\frac{29}{30} P_0$.

Consider now what happens if some disturbance initiates an upward movement of air. The air will expand and in the course of this it will cool. Assuming the cooling to be purely adiabatic, ground temperature to be T_0, and the temperature at the 1000-ft level to be T, we have from equation **(4.24)**

$$\frac{T}{T_0} = \left(\frac{P}{P_0}\right)^{\frac{\gamma-1}{\gamma}} = \left(\frac{P}{P_0}\right)^{\frac{1\cdot4-1}{1\cdot4}} = \left(1-\frac{1}{30}\right)^{\frac{2}{7}} = \left(1-\frac{2}{210}\cdots\right),$$

(4.25c)

since γ for air is approximately $1\cdot4$. If $T_0 = 300°C$.

$$T = 300\left(1-\frac{2}{210}\right) = 300-3,$$

so that the temperature drop is 3°C. This is close to the observed temperature drop per 1000 ft. In practice if there is an appreciable amount of moisture present in the air, cooling leads to condensation and this releases energy which lessens the cooling. A more representative value is about 2°C. per 1000 ft.

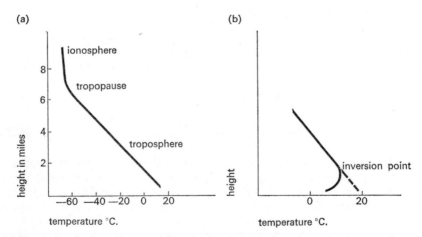

Figure 10. The decrease of temperature with height in air. (*a*) steady conditions, (*b*) early morning.

Figure 10(*a*) shows the variation of temperature with height. During the day the temperature drops from about 20°C. at ground-level to −60°C. at a

40 Gases, liquids and solids

height of 7 or 8 miles, where the troposphere ends. Often in the early morning the temperature gradient is reversed near the ground level due to the cooling of the earth overnight [figure 10(*b*)]. Inversion temperatures of this type can be an important cause of fog and smog.

4·2 Elementary kinetic theory of ideal gases

The idea that gas molecules are in random motion was considered by the Greeks [as Lucretius' description of Brownian motion shows (see the end of Chapter 5)]. It was revived in a surprisingly modern form by D. Bernouilli (1700–1782) in 1738, but seems not to have been widely adopted until J. C. Maxwell (1831–1879) took the matter in hand and provided practically the whole of our present theory of gases. We first present a very simple theory. Its basic assumptions are:

    (i) the gas consists of identical molecules of mass m;
    (ii) they have zero size, and do not collide with one another;
    (iii) they exert no forces on one another;
    (iv) they undergo random motion within the gas and their collisions with the walls of the container are perfectly elastic.

We now consider the collisions of the gas molecules with the container walls and show that the pressure exerted by the gas on the walls is due to the momentum transfer accompanying the collisions.

Suppose there are n molecules per c.c. They all have varying velocities but because the numbers are generally so enormous we can subdivide these up into groups of $n_1, n_2, n_3 \ldots$ molecules per c.c. with velocity ranges c_1 to $c_1 + dc$; c_2 to $c_2 + dc$; c_3 to $c_3 + dc$, etc.

Let the gas be held in a cubic container of side l, and let us first consider the group with the velocity c_1 – there are $n_1 l^3$ of these in the container.

We resolve c_1 into three mutually orthogonal components u_1, v_1, w_1, parallel to the sides of the cube. Clearly

$$u_1^2 + v_1^2 + w_1^2 = c_1^2. \tag{4.26}$$

Along the u_1 direction the molecule has momentum mu_1 normal to the face of the cube before collision and momentum $-mu_1$ after collision. Thus for each collision

$$\text{momentum transfer to wall} = 2mu_1. \tag{4.27}$$

The molecule has to travel across the cube and back again (i.e. a distance of $2l$) to make the next collision. The time taken is $\dfrac{2l}{u_1}$.

Hence the molecule makes $\dfrac{u_1}{2l}$ collisions per second.   **(4.28)**

41   Perfect gases – bulk properties and simple theory

The momentum transfer per molecule per second is the product of equations (4.27) and (4.28), i.e.

$$2mu_1 \left(\frac{u_1}{2l}\right) = \frac{mu_1^2}{l}.$$ (4.29)

Consequently for all the molecules of this group the total momentum transfer per second is

$$\sum \frac{m}{l} u_1^2 = \frac{m}{l} \sum u_1^2.$$ (4.30)

We now define the average value of u_1^2 for all this group as

$$\overline{u_1^2} = \frac{\sum u_1^2}{n_1 l^3},$$ (4.31)

so that equation (4.30) becomes

$$\frac{m}{l} n_1 l^3 \overline{u_1^2} = mn_1 \overline{u_1^2} l^2.$$ (4.32)

The momentum transfer per second is the force and this is exerted on a face of area l^2. These molecules therefore exert a pressure p on the wall of amount

$$p = mn_1 \overline{u_1^2}.$$ (4.33)

We now make the following observation. For this group of molecules the velocity c_1 is the same but their paths are uniformly distributed in all directions; when averaged over the large number of molecules involved there can be no preferred direction so that

$$\overline{u_1^2} = \overline{v_1^2} = \overline{w_1^2} = \tfrac{1}{3}c_1^2.$$ (4.34)

Equation (4.33) becomes

$$p = \tfrac{1}{3}mn_1 c_1^2.$$ (4.35)

For all the groups of molecules the total pressure becomes

$$p = \sum \tfrac{1}{3}mn_i c_i^2$$
$$= \tfrac{2}{3} \sum \tfrac{1}{2}mn_i c_i^2 = \tfrac{2}{3} \text{ (total kinetic energy per c.c.).}$$ (4.36)

For the total volume V we have

$$PV = \tfrac{2}{3} \text{ (total kinetic energy of molecules in the volume } V\text{).}$$ (4.37)

We now define the mean square velocity of all the molecules in the gas as

$$\overline{c^2} = \frac{n_1 c_1^2 + n_2 c_2^2 + n_3 c_3^2 \ldots}{n_1 + n_2 + \ldots}$$
$$= \frac{\sum n_i c_i^2}{n}.$$ (4.38)

42   Gases, liquids and solids

We can insert this in equation (4.36) and obtain

$$P = \tfrac{1}{3} \Sigma mn_i \, c_i^2 = \tfrac{1}{3}mn\overline{c^2}. \qquad \text{(4.39)}$$

For a gas occupying volume V, where the number of molecules is N where $N = nV$ we have

$$PV = \tfrac{1}{3}mN\overline{c^2}. \qquad \text{(4.40)}$$

For a gram-molecule of a gas N is Avogadro's number, N_0, and

$$PV = \tfrac{1}{3}mN_0\overline{c^2} = RT \qquad \text{(4.41)}$$

Order of magnitude of $(\overline{c^2})^{\frac{1}{2}}$. The square root of $\overline{c^2}$ is called the root mean square velocity. As we shall see below it is a little less than the arithmetic mean of the velocities but not very different. From equation (4.41) $RT = \tfrac{1}{3}mN_0\overline{c^2} = \tfrac{1}{3}M\overline{c^2}$ where M is the molecular weight. Consequently

$$\overline{c^2} = \frac{3RT}{M}. \qquad \text{(4.42)}$$

Typical values for $(\overline{c^2})^{\frac{1}{2}}$ are given in Table 10 for gases at 0°C. ($T = 273$°K.).

Table 10   *Typical values of r.m.s. velocity*

Gas	r.m.s. velocity m. sec.$^{-1}$
H_2	1840
He	1300
O_2	650
Ar	410
Benzene vapour	290
Mercury vapour	180
Electron gas	100,000

4·3   The ether theory of the ideal gas

Before congratulating ourselves on the validity of the kinetic theory we may consider one of the theories which preceded it, and I am much indebted to Professor Eric Mendoza for pointing it out to me. This theory was based on the Newtonian view that a gas was like a solid. In a solid the molecules are held together by attractive forces. In a gas they are repelled by some agency which pushes them apart but still leaves them in a, more or less, uniform array. What pushes them apart? One view was that it was the swirling ether.

If each molecule is surrounded by a shell of ether of radius r rotating at a very high speed, the centrifugal force exerts a pressure on the neighbouring sphere and so keeps the molecules apart. It is as though the molecules are surrounded by balloons which press against one another. If the ether has a mass m and the velocity is somehow in all directions, but confined to the great circles of the sphere, the net centrifugal force is $\dfrac{mv^2}{r}$ and it is exerted over an area $4\pi r^2$. The pressure exerted on the balloon shell is therefore

$$p = \frac{\dfrac{mv^2}{r}}{4\pi r^2} = \frac{mv^2}{4\pi r^3} = \frac{\tfrac{1}{3}mv^2}{\dfrac{4\pi r^3}{3}}.$$

The numerator on the right-hand side is $\tfrac{2}{3}$ times the kinetic energy of the ether; the denominator is the volume occupied by the ethereal shell. We are left with the relation

$$p = \tfrac{2}{3}\text{ (kinetic energy of ether per unit volume).}$$

This is very similar to equation (4.36) and shows that the right answer does not necessarily guarantee the validity of the physical model, a most salutory warning. There is, of course, more to this than coincidence. Any gas theory which attributes the gas pressure to the kinetic energy in the gas is bound to give a relation similar to that obtained in the kinetic theory. This follows because pressure has the dimensions of $\dfrac{\text{force}}{\text{area}}$ and energy per unit volume has the dimensions of $\dfrac{(\text{force} \times \text{distance})}{\text{volume}}$ which is also $\dfrac{\text{force}}{\text{area}}$.

4·4 Some deductions from kinetic theory

4·4·1 *Thermal equilibrium between gas and container*

We now make use of a treatment described by J. Jeans in his *Kinetic Theory of Gases* to analyse the conditions which determine the thermal equilibrium which must exist between the molecules of a gas and the molecules of the container. We start off by considering the collision between one gas molecule and one wall molecule. Let m, M be the mass of the gas molecule and wall molecule respectively, and let their velocity components in the x, y, z directions be u, v, w and U, V, W respectively. Suppose the collision is along the direction of u, U, so that v, w, and V, W, are unchanged. After collision u and U become u' and U'.

For an elastic collision the relative velocity is reversed so that

$$U' - u' = -(U - u). \tag{4.43}$$

44 Gases, liquids and solids

The momentum is conserved, so that

$$mu + MU = mu' + MU'$$

or

$$m[u - u'] = -M[U - U'].$$ (4.44)

Combining equations (4.43) and (4.44) we have

$$U' = \frac{1}{m+M}[(M-m)U + 2mu].$$

The gain in kinetic energy of the wall molecule as a result of the collision is

$$\frac{1}{2}M(U'^2 - U^2) = \frac{M}{2}(U' + U)(U' - U)$$

$$= \frac{2Mm}{(m+M)^2}[mu^2 - MU^2 + (M-m)uU].$$ (4.45)

If we now consider the average behaviour of a large number of collisions we note that u must always be positive if collision is to occur, whereas U may be positive or negative and its average value must be zero. Consequently for a large number of collisions between gas molecules and wall molecules, the average value of uU is zero and the average gain of energy of each wall molecule becomes

$$\frac{2Mm}{(m+M)^2}[\overline{mu^2} - \overline{MU^2}],$$ (4.46)

where barred symbols refer to the mean square velocities of u and U. For thermal equilibrium there can be no net transfer of energy to the wall. Hence

$$\overline{mu^2} = \overline{MU^2}.$$ (4.47)

Similar arguments may be applied to the other components (v, V) and (w, W). The final conclusion is that for thermal equilibrium

$$\overline{mc^2} = \overline{MC^2},$$ (4.48)

where $\overline{c^2}$ and $\overline{C^2}$ are the mean square resultant velocities of the gas and wall molecules respectively. This equation implies that for thermal equilibrium the colliding molecules must have the same average kinetic energy.

4·4·2 Boyle's law

For a given quantity of gas

$$PV = \tfrac{2}{3}(\text{total K.E. in volume } V).$$

Since the kinetic energy is constant if the temperature is constant this implies that

$$PV = \text{constant}.$$

45   Perfect gases – bulk properties and simple theory

4·4·3 Charles' law

If the pressure is kept constant the volume increases linearly with the kinetic energy of the molecules. Only a sophisticated thermodynamic treatment can show that the kinetic energy itself is proportional to the absolute temperature.

4·4·4 Avogadro's law

Two different gases at the same temperature must have the same mean molecular kinetic energy, since they are each individually in equilibrium with the walls of their container. But

$$PV = \tfrac{2}{3} \text{ (total K.E. in volume } V\text{)}$$

$$= \tfrac{2}{3} \text{ (average K.E. per molecule} \times \text{number of molecules in volume } V\text{).}$$

Hence (at constant temperature) PV will have the same value for all gases provided they contain the same number of molecules.

4·4·5 Dalton's law of partial pressure

If we have two gases which separately have average molecular kinetic energies E_1 and E_2, and a concentration of n_1 and n_2 per c.c. respectively we may write,

$$P_1 = \tfrac{2}{3} \text{ (K.E. per unit volume)}$$

$$P_1 = \tfrac{2}{3} (E_1 n_1).$$

Similarly

$$P_2 = \tfrac{2}{3} (E_2 n_2).$$

For gases at equal temperatures $E_1 = E_2$, so by adding the gases together a new concentration $n = n_1 + n_2$ is obtained and the combined pressure is

$$P = \tfrac{2}{3} E n = \tfrac{2}{3} E(n_1 + n_2) = \tfrac{2}{3} n_1 E_1 + \tfrac{2}{3} n_2 E_2$$

$$= P_1 + P_2. \tag{4.49}$$

4·4·6 Mean free path

We consider here the average distance between collisions of the gas molecules Suppose σ is the effective diameter of a molecule, then any other molecule whose centre is within a distance σ from the molecule considered will touch it [see figure 11(a)]. For our simplest model we assume all the other gas molecules to be instantaneously at rest and follow the fate of the single mobile molecule. If this molecule has, in a given instant of time, travelled a distance l, it will have swept out a volume $\pi\sigma^2 l$ within which any other

46 Gases, liquids and solids

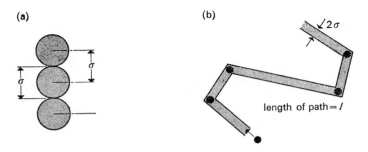

Figure 11. (*a*) The effective diameter σ of a molecule. (*b*) A single molecule is allowed to travel through gas molecules considered instantaneously at rest.

molecule will touch it [see figure 11(*b*)]. Thus if there are n molecules per c.c. the number of molecules with which our moving molecule will collide will be $\pi\sigma^2 ln$ molecules.

The mean distance between molecular collisions λ may now be defined as

$\dfrac{\text{(distance travelled)}}{\text{(number of collisions)}}$, that is

$$\lambda = \frac{l}{\pi\sigma^2 ln} = \frac{1}{\pi\sigma^2 n}. \tag{4.50}$$

This is, of course, the simplest approach possible. In practice all the molecules are not immobilized. If we allow for the true relative velocity between the molecules this reduces λ by a factor $1/\sqrt{2}$. Again in a collision molecules will only rarely have equi-velocity head-on collisions and bounce away from one another; usually there will be some persistence of velocity in the original direction after collision and this will complicate the analysis. Maxwell also considered the influence of intermolecular forces. All these refinements merely change the value of λ by a small numerical factor and we shall not consider this further. Some idea of the complexity of the problem may be gathered from Jeans' book on the kinetic theory of gases.

A more important question is: how far does the concept of molecular collisions and molecular diameter affect our simple kinetic theory, where the basic assumption made was that the molecules have no size, and do not collide with one another in travelling across the container from one wall to the other? The answer is that the finite size of the molecules does influence the simple theory since the 'free' volume available for molecular movement is reduced. This will be discussed in greater detail when we consider the behaviour of 'imperfect' gases in Chapter 6. The collisions between the molecules do not, however, alter the simple theory. As we shall see in the next chapter collisions deflect molecules and remove them from the group under consideration, but these are constantly replaced (as a result of other collisions) by molecules of other groups. Thus the behaviour is the same as though we could divide the molecules into non-colliding groups with specified velocities and directions.

We see from equation (4.50) that λ varies as $1/n$, that is, it varies as the reciprocal of the pressure P. Taking an average value of $\sigma = 3$ Å., we find that for air at s.t.p. $\lambda \simeq 1000$ Å., whilst the mean distance between molecules is of the order $(1/n)^{\frac{1}{3}} \simeq 30$ Å. These values 3, 30, 1000 Å. for diameter, separation and mean free path are typical of gases under normal conditions of temperature and pressure.

4·4·7 Softening of molecules

Molecules are not hard impenetrable spheres. If the molecular speed increases they will, during collision, penetrate further into the repulsion fields of their neighbours; but this effect is very small since the repulsive forces are very powerful.

A more important effect is that which arises from the attractive or repulsive forces themselves. These will deflect the molecular paths and produce a marked change in the effective collision diameter of the molecules. An analysis for repulsive forces, falling off as the fifth power of the separation, was first given in analytical form by Maxwell. Forces of this type are not representative of most gas molecules. A more realistic model is one which considers the attractive forces which molecules exert on others passing nearby. This is illustrated schematically in figure 12. If the attractive forces are zero [figure 12(a)], molecule B will just hit A if the separation is less than σ_0. If, however, attractive forces are not negligible B may well hit A even though its original direction is such that it would miss it [figures 12(b) (c)].

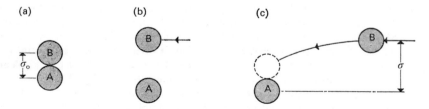

Figure 12.   The effect of attractive forces on the effective collision diameter.

Hence although there is no change in the size of the molecule, from the point of view of collision and transport phenomena, it behaves as though it had a much larger effective cross-section σ. Thus slow molecules will tend to collide even if they are very far apart, i.e. σ tends to infinity as T tends to 0. On the other hand, for very high velocities, i.e. very high temperatures, there will be practically no deflection of the molecular paths and collision will only occur when the separation is close to σ_0. A useful empirical relation for the effective molecular diameter σ at temperature T in terms of the 'geometric' diameter σ_0 is

$$\sigma^2 = \sigma_0^2 \left(1 + \frac{C}{T}\right), \tag{4.51}$$

where C is known as Sutherland's constant for the gas. Some typical values of Sutherland's constant are given in Table 11a, and Table 11b gives values of the effective cross-section of a number of gases at temperatures from $-100°C$ to $300°C$. The effect is not large but is not negligible especially for larger molecules.

Table 11a *Typical values of Sutherland's constant C*

Gas	Ne	He	O_2	Cl_2	Steam
C °K.	56	72	125	350	650

Table 11b *Effective cross-section compared with cross-section at $0°C$.*

(Based on values of C given in G. W. C. Kaye and T. H. Laby (1959), *Tables of Physical and Chemical Constants*, 12th edn, Wiley).

Gas	$-100°$	$0°$	$100°$	$200°$	$300°$
Ne	1·1	1	0·94	0·92	0·9
H_2	1·13	1	0·94	0·92	0·9
O_2	1·18	1	0·9	0·87	0·81
Cl_2	1·32	1	0·85	0·76	0·70
H_2O	—	1	0·79	0·68	0·61

4·4·8 *Number of molecular collisions per unit area per second*

Consider a surface of area A subjected to molecular collisions. Divide the molecules into groups n_1 per c.c. with velocity c_1; n_2 per c.c. with velocity c_2, etc. Consider the first group n_1 and assume that they are all moving normal to the surface in question. Then in one second all the molecules in the volume $c_1 A$ will strike the surface, i.e. $n_1 c_1 A$ molecules will strike. Hence the number of collisions per sq. cm. per second would be $n_1 c_1$.

We now make use of the concept known as the Joule classification, which we used (without naming it) in our treatment of the kinetic theory. The velocity

49 **Perfect gases – bulk properties and simple theory**

c_1 of a molecule may be described in terms of its u_1, v_1, w_1, components. Since for any one group all directions are equally possible the number, on average, travelling in any *one* direction of given sign will be one sixth of the total number.

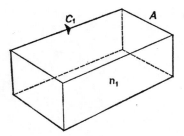

Figure 13. Collision of molecules travelling with velocity c_1 normal to a surface of area A.

Hence the collision rate cm.$^{-2}$ = $\frac{1}{6}n_1 c_1$.
For all the groups we have

$$\text{collisions cm.}^{-2}\text{ sec.}^{-1} = \tfrac{1}{6}\Sigma n_1 c_1 = \tfrac{1}{6}n\bar{c} \tag{4.52}$$

where n is the total number per c.c. and $\bar{c}$ is the mean velocity.

4·5 Transport phenomena

There are three bulk properties of gases which can be explained satisfactorily in terms of a simple molecular model, involving the transport of properties through the gas. For this reason they are called transport phenomena. Viscosity can be explained in terms of transport of momentum, heat conduction in terms of transport of thermal energy, diffusion in terms of molecular transport produced by concentration gradients.

4·5·1 *Viscosity*

We consider the simplest arrangement where a plane surface YY is at rest and another parallel surface ZZ at a distance h away moves to the right with uniform velocity u. It is then found that a force F must be applied to the top plate to maintain the velocity u against the viscous drag of the fluid between the plates (figure 13). If the flow of the fluid between the surfaces is along lines parallel to the planes YY and ZZ (lamellar flow) it is found that F is proportional to the area A of the moving surface and the velocity gradient u/h. In the more general case where the velocity gradient is not uniform across the gap, F is proportional to A and to du/dy.

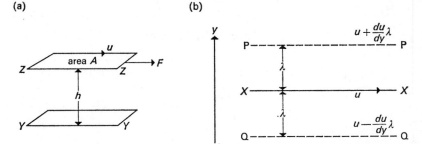

Figure 14. (a) The concept of viscosity. (b) Model for describing transport of momentum in molecular terms.

The factor of proportionality η is introduced such that

$$F = \eta A \frac{du}{dy}, \tag{4.53}$$

where η is the viscosity of the liquid or gas between the surfaces.

We now show that the viscosity of a gas can be expressed in terms of momentum transported across the gas. Consider a plane XX within the gas parallel to the direction of the stream lines [figure 14(b)]. Let the velocity of flow at XX be u and the velocity gradient du/dy. We consider the molecules entering and leaving this plane. We assume

(a) that the flow velocity u is very small compared with the mean gas velocity $\bar{c}$;

(b) that the only molecules reaching XX are those, which, on an average, have just made their last collision at a distance λ from XX. Then at P the molecules have a flow velocity $u + \frac{du}{dy}\lambda$ and at Q a flow velocity $u - \frac{du}{dy}\lambda$.

From assumption (a) the number of molecules crossing an area A in one second is $\frac{n\bar{c}}{6}A$. From Q these molecules bring to XX horizontal momentum

$$m\left(u - \frac{du}{dy}\lambda\right)\frac{n\bar{c}}{6}A. \tag{4.54}$$

From P a similar flow brings horizontal momentum

$$m\left(u + \frac{du}{dy}\lambda\right)\frac{n\bar{c}}{6}A. \tag{4.55}$$

The plane XX itself is also discharging $\dfrac{n\bar{c}A}{6}$ molecules per second towards both P and Q. The resultant behaviour is summarized in Table 12.

Table 12 *Molecules entering and leaving an area* A *in plane XX per sec.*

	Number	Horizontal momentum
Entering from P	$\frac{1}{6}n\bar{c}A$	$m\left(u+\dfrac{du}{dy}\lambda\right)\dfrac{n\bar{c}A}{6}$
Entering from Q	$\frac{1}{6}n\bar{c}A$	$m\left(u-\dfrac{du}{dy}\lambda\right)\dfrac{n\bar{c}A}{6}$
Leaving from XX upwards	$\frac{1}{6}n\bar{c}A$	$mu\,\dfrac{n\bar{c}A}{6}$
Leaving from XX downwards	$\frac{1}{6}n\bar{c}A$	$mu\,\dfrac{n\bar{c}A}{6}$
Net addition to XX	0	0

We see that there is no accumulation of molecules or momentum in plane XX. There is, however, a transport of momentum per second *through* the plane given by the difference between equations **(4.55)** and **(4.54)**. This amounts to

$$\frac{n\bar{c}A}{6}\left[m\left(u+\frac{du}{dy}\lambda\right)-m\left(u-\frac{du}{dy}\lambda\right)\right]$$

$$=\frac{m\lambda n\bar{c}}{3}A\frac{du}{dy}. \tag{4.56}$$

This horizontal momentum per second is a horizontal force which is transmitted through the gas to the top moving surface. As we saw before, in equation **(4.53)**, the tangential force on the top plate is

$$F = \eta A\frac{du}{dy}.$$

By comparison we see that we have derived a relation for the viscosity of the gas

$$\eta = \tfrac{1}{3}m\lambda n\bar{c} \tag{4.56a}$$

or $\quad\quad \eta = \tfrac{1}{3}\dfrac{m}{\pi\sigma^2}\,\bar{c}. \tag{4.56b}$

We conclude that the viscosity of a gas is proportional to $\bar{c}$, that is to $T^{\frac{1}{2}}$ and is independent of n, that is, of the pressure. Both these conclusions are well supported by experiment. If however the pressure is too low and the mean free path becomes comparable with the distance between the moving surfaces, equations (4.56) are no longer reliable. This is discussed in the following paragraphs, which may be omitted on a first reading.

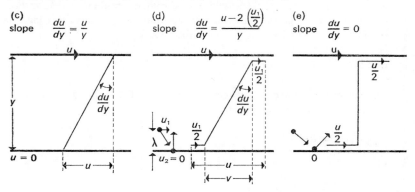

(c) slope $\dfrac{du}{dy} = \dfrac{u}{y}$ (d) slope $\dfrac{du}{dy} = \dfrac{u - 2\left(\dfrac{u_1}{2}\right)}{y}$ (e) slope $\dfrac{du}{dy} = 0$

Figure 14. (c) Ignoring the problem of the solid boundaries the velocity gradient across the bulk of the gas is $\dfrac{u}{y}$. (d) If impinging molecules from the last mean-free-path separation acquire the tangential velocity of the solid wall there is an effective discontinuity in the mean tangential gas velocity close to the wall. This leads to a reduction in the velocity gradient in the bulk of the gas. (e) If impinging molecules rebound 'elastically' so that their horizontal velocity component after impact is the same as before, no horizontal momentum is communicated to the walls. There is complete slipping of the molecules over the walls and zero velocity gradient in the bulk of the gas.

Suppose the distance between the plates is y, the lower plate is stationary and the upper plate has velocity u. If y is very large compared with the mean free path λ, the velocity gradient across the gas [see figure 14(c)] is

$$\frac{du}{dy} = \frac{u}{y}. \tag{4.57a}$$

Let us now consider in greater detail the situation at the solid plates themselves. It is clear from our derivation of η that at the solid walls only one half of the transport process is operative since molecules cannot pass *through* the walls. Some change in flow rates must occur close to the walls if the appropriate shear stress is to be communicated through the gas to the solid wall.

Consider first the stationary wall. Molecules arriving after their last collision at a mean distance λ away, bring with them horizontal velocity u_1. If all the molecules striking the wall are absorbed and then re-emitted with the tangential velocity of the (stationary) wall, this tangential velocity u_2 is zero

53 Perfect gases – bulk properties and simple theory

[see figure 14(d)]. Thus the average velocity of the molecules very close to the stationary wall is the average of u_1 and zero, that is $u_1/2$. This is equivalent to saying that the molecules slip over the walls with velocity $u_1/2$. A similar slip occurs at the moving surface. Between these two planes at which slip occurs there is full transport of molecules to and from the plates, so that a constant velocity gradient may be maintained throughout the bulk of the gas in accordance with equation (4.57a). The velocity gradient is determined by the fact that between the stationary surface and the first mean free path there is a change in velocity of $u_1/2$ over a distance λ,

that is $\quad \dfrac{u_1}{2} = \lambda \dfrac{du}{dy}$.

The velocity difference v between the gas near the stationary wall and the gas near the moving wall [see figure 14(b)] is clearly

$$v = u - \frac{2u_1}{2},$$

or

$$v = u - 2\lambda \frac{du}{dy}.$$

Over this region there is a uniform velocity gradient of amount

$$\frac{v}{y} = \frac{\left(u - 2\lambda \dfrac{du}{dy} \right)}{y} = \frac{du}{dy},$$

or

$$\frac{du}{dy} = \frac{u}{y + 2\lambda}. \tag{4.57b}$$

We now see that the shear stress across every section of the bulk of the gas is communicated to the solid walls by the velocity jump close to the solid surface. Comparing equation (4.57b) with (4.57a), the apparent viscosity of the gas is reduced by a factor

$$\frac{y}{y + 2\lambda}. \tag{4.57c}$$

Clearly if the mean free path is an appreciable fraction of the separation there will be a large reduction in the apparent viscosity. In the limit if λ is greater than y, none of the concepts used in this analysis is applicable.

There is yet a further factor involved. The molecules striking the surface may not all take up the tangential velocity of the surfaces before rebounding. In the extreme case, if the tangential velocities of the molecules leaving the surface are the same as those striking it, there will be no transfer of tangential momentum to the walls. Consequently there will be no resultant velocity gradient in the bulk of the gas, but a velocity discontinuity of $u/2$ at a short distance from each of the walls [figure 14(e)]. This corresponds to perfect 'slipping' of the gas over the wall surfaces without the transfer of any shear stress. Under these conditions the gas would appear to have zero viscosity. In practice the situation is much closer to that involving adsorption and re-emission with the wall velocity. As a result the simple correction, given by equation (4.57c), is valid.

Finally we may refer the reader to a more quantitative treatment of this problem given by S. Chapman and T. G. Cowling in *The Mathematical Theory of Non-Uniform Gases*, Cambridge University Press, 1939, chapter 6.* If θ is the fraction of molecules absorbed and emitted with the tangential velocity of the wall, then the fraction reflected with the same tangential velocity as the incident molecules is $1-\theta$. The average tangential velocity of the 'reflected' molecules, for example from the stationary wall, is

$$u_2 = (1-\theta)\, u. \qquad (4.57d)$$

The mean velocity at the wall is $\frac{1}{2}(u_1+u_2)$. This is the velocity of slip. Hence

$$u_1 - \tfrac{1}{2}(u_1 + u_2) = \lambda\frac{du}{dy}. \qquad (4.57e)$$

Using equation (14.57d) this gives for the slip velocity

$$\tfrac{1}{2}(u_1+u_2) = \frac{2-\theta}{\theta}\,\lambda\frac{du}{dy}. \qquad (4.57f)$$

Because of similar slip at the moving wall the velocity difference across the bulk of the gas is

$$u-2\left(\frac{2-\theta}{\theta}\right)\lambda\frac{du}{dy}. \qquad (4.57g)$$

The apparent viscosity of the gas is thus reduced by the factor

$$\frac{y}{y+2\lambda\left(\dfrac{2-\theta}{\theta}\right)}. \qquad (4.57h)$$

For $\theta = 1$ this reduces to equation (4.57c); for $\theta = 0$, the ratio is infinite. This corresponds to complete slip at the walls as shown in figure 14(e).

4·5·2 *Thermal conductivity*

Consider heat flow across the material lying between parallel surfaces YY and ZZ [figure 15(a)]. If the temperature of the hotter surface is T_1 and the

* I am indebted to Mr N. J. Holloway and Mr B. Scruton for bringing this to my attention.

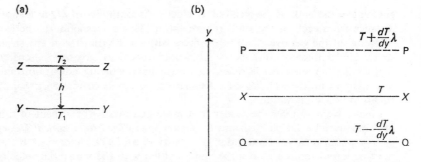

(a)   (b)

Figure 15. (a) The concept of thermal conductivity. (b) Model for describing transport of thermal energy in molecular terms.

colder T_2, the rate of flow of heat is proportional to $\dfrac{T_1 - T_2}{h}$ and to the area of the surfaces. More generally we write for the heat flow per second

$$Q = -KA\frac{dT}{dy},\tag{4.58}$$

where K is the thermal conductivity of the material and dT/dy the temperature gradient. The negative sign is because there is a flow of heat from the hotter to the colder region (positive Q for negative dT/dy).

We again assume that in a gas heat conduction is due to the transport of thermal energy. If c_V is the specific heat at constant volume per molecule the rate of transport of thermal energy into plane XX from planes at a distance λ from it is:

from the upper plane:

$$\frac{n\bar{c}A}{6}\,c_V\left(T+\frac{dT}{dy}\lambda\right)\tag{4.59a}$$

from the lower plane:

$$\frac{n\bar{c}A}{6}\,c_V\left(T-\frac{dT}{dy}\lambda\right).\tag{4.59b}$$

Again we may show that there is no net accumulation of molecules or of thermal energy in plane XX. There is, however, a net transport of thermal energy per second of amount equal to the difference between equations (4.59a) and (4.59b).

Hence

$$Q = -\frac{n\bar{c}\lambda}{3}\,c_V \times A\frac{dT}{dy}.$$

56   Gases, liquids and solids

By comparison with equation (4.58) we see that the thermal conductivity is given by

$$K = \frac{n\bar{c}\lambda}{3}\, c_V.\qquad\qquad (4.60)$$

As in the preceding discussion of viscosity the temperature gradient near solid walls will depend on the degree of thermal equilibration between the temperature of the molecules striking the walls, and the walls themselves. If the molecules acquire the temperature of the walls, the temperature very close to the wall will show a small discontinuity, and there will be a reduction in the temperature gradient through the bulk of the gas. If the collisions are perfectly elastic so that the molecules strike the wall and are reflected with their original thermal velocity, that is they ignore the temperature of the wall, no heat can be transferred to or from the walls. There will be a large discontinuity in temperature at the walls: the total temperature drop will occur in a very short distance comparable with λ and there will be a negligible temperature gradient across the mass of the gas. Consequently there will be no heat transfer. This is the limiting case in which the 'heat transfer coefficient' is zero.

Ratio of thermal conductivity and viscosity. From equations (4.60) and (4.56a) we have

$$\frac{K}{\eta} = \frac{c_V}{m} = \frac{c_V N_0}{m N_0} = \frac{C_V}{M}$$

where C_V is the specific heat per gram-molecule and M the molecular weight. We therefore expect to find

$$\frac{K}{\eta}\frac{M}{C_V} = 1.\qquad\qquad (4.61)$$

The experimental value of this ratio lies between 1·4 and 2·5 for a very wide range of gases. The discrepancy is due partly to molecular repulsions which tend to reduce η. A more important factor is that, in thermal conduction, the transport of the translational kinetic energy takes place more efficiently than the transfer of the rotational and vibrational forms of thermal energy. This leads to an increase in K relative to η, and more sophisticated theories are available that agree well with the experimental ratio of $\frac{K}{\eta}\frac{M}{C_V}$. Even the order of agreement with the simple theory provides strong support for the essential validity of the transport mechanisms described above.

4·5·3   *Self diffusion*

If we have a concentration gradient in a gas, the molecules will move from the more concentrated to the less concentrated regions via molecular collisions;

this involves a kinetic diffusion process. If over a distance dx the concentration increases by dn molecules per c.c. the concentration gradient is dn/dx. It is then found that the number of molecules per second crossing an area A normal to the gradient, i.e. the rate of diffusion of molecules per second, can be written

$$\frac{dN}{dt} = -D\frac{dn}{dx}A,$$

where D is called the coefficient of self diffusion and the negative sign implies flow in the direction of smaller concentration.

Consider again a plane XX where the concentration is n and neighbouring planes P and Q distant from it where the concentrations are $n+\dfrac{dn}{dx}$ and $n-\dfrac{dn}{dx}$ respectively. The number of molecules per second crossing from P

$$= \frac{1}{6}\left(n+\frac{dn}{dx}\lambda\right)\bar{c}A \qquad\qquad (4.62a)$$

and from Q

$$= \frac{1}{6}\left(n-\frac{dn}{dx}\lambda\right)\bar{c}A \qquad\qquad (4.62b)$$

There are also molecules leaving on each side of XX of amount $\frac{1}{6}n\bar{c}A$. Again there is no net accumulation of molecules in XX, but the net transfer is

$$-\frac{\lambda\bar{c}}{3}\frac{dn}{dx}A = -D\frac{dn}{dx}A.$$

Consequently

$$D = \frac{\lambda\bar{c}}{3}$$

or $\qquad D = \dfrac{\bar{c}}{3\pi n\sigma^2}.$ $\qquad\qquad\qquad\qquad (4.63)$

It is interesting to see the orders of magnitude involved in D. For air at s.t.p. we have:

$$\sigma = 2\cdot5\text{ Å.}; \quad \lambda = 1000\text{ Å.}; \quad \bar{c} = 45000\text{ cm. sec.}^{-1}$$

This gives a value of D of order of 1 molecule per sq. cm. per unit concentration gradient. D is very small compared with $\bar{c}$ due to the retarding effect of molecular collisions.

There is only one direct way of studying D. This is to incorporate some radioactive molecules in one specimen of gas and place it in contact with an identical gas containing no radioactive molecules. The diffusion of the radioactive species may then be followed, say, using a Geiger counter. Even here there is some difficulty since the radioactive species may not be the same size as the non-radioactive. For this reason most diffusion experiments involve the diffusion of one gas through another. If the concentration of gas 1 is n_1 per c.c. and that of gas 2 is n_2 per c.c.; if c_1 and c_2 are the respective molecular velocities and σ_1 and σ_2 the respective molecular cross-sections, the diffusion coefficient of gas 1 into gas 2 may be shown to be given by

$$D_{1,2} = \frac{\sqrt{(c_1^2 + c_2^2)}}{3\pi(n_1 + n_2)(\sigma_1 + \sigma_2)^2}. \tag{4.64}$$

A typical example of $D_{1,2}$ is the diffusion of a vapour through air. Consider a cylindrical vessel of cross-sectional area A (figure 16). It contains a small

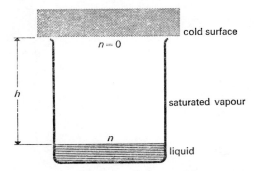

Figure 16. Diffusion of molecules from the surface of a liquid where vapour concentration corresponds to that of saturated vapour. The vapour condenses on a cold surface at the top of the vessel so that at this region the vapour concentration is virtually zero.

quantity of liquid at the bottom, and at the top open end is a cold surface on which the vapour immediately condenses. The vessel is full of air and diffusing vapour. Just above the liquid the vapour is at its saturation pressure implying a concentration n molecules per c.c. At the cold surface the vapour pressure is almost zero. In the steady state there can be no accumulation of vapour molecules in any section. This implies that the diffusion rate across every section is a constant. Consequently there must be a linear concentration drop from n at the surface of the liquid to zero at the cold surface, distance h.

Then $\dfrac{dn}{dx} = \dfrac{n}{h}$.

Hence the mass of vapour transferred per second and condensed on the cold surface is

$$w = D_{1,2}\, A\, \frac{dn}{dx}\, m = D_{1,2}\, A\, \frac{n}{h}\, m,$$

where m is the weight of the molecule. The product nm is the density ρ of the saturated vapour. If we, therefore, determine the weight w of liquid condensed on the cold surface (or lost from the liquid) per second, we have

$$D_{1,2} = \frac{wh}{A\rho}. \tag{4.65}$$

In this way it is possible to deduce values of $(\sigma_1 + \sigma_2)^2$. Some typical values of σ deduced from measurements of η, k and D are given in Table 13.

Table 13 *Molecular diameters σ $(10^{-8}$ cm.)*

Gas	Deduced from		
	η	K	D
Ar	3·6	3·2	3·8
CO_2	4·5	3·8	4·5
CO	3·7	3·7	3·8
Cl_2	5·4	5·9	—
He	2·1	1·9	—
H_2	2·7	2·7	2·7
CH_4	4·1	4·3	4·0
N_2	3·7	3·6	3·7
O_2	3·6	3·5	3·6

(Adapted from G. W. C. Kaye and T. H. Laby (1959), *Tables of Physical and Chemical Constants,* 12th edn, Wiley.)

4·6   Sound waves in a gas

4·6·1   *Velocity of sound in a gas*

All elastic media can transmit waves of well-defined velocity and frequency, and with solids the waves can involve shear as well as compression. But with gases the only waves that can be transmitted are those involving a succession of compressions and rarefactions. The velocity of such a longitudinal wave is given by

$$v = \sqrt{\frac{\text{elastic modulus}}{\text{density}}}. \tag{4.66}$$

The elastic modulus is the bulk modulus defined as the pressure increment dP

divided by the fractional volume increment dV/V. Since a pressure increase produces a diminution in volume it is usual to define the modulus as

$$-\frac{dP}{\left(\dfrac{dV}{V}\right)}.$$

We shall soon show that sound waves are adiabatic and not isothermal. However, at this point, we shall assume isothermal waves. If we write

$$PV = RT$$

then for an isothermal change $PdV + VdP = RdT = 0$, so that the modulus

$$-\frac{dP}{\left(\dfrac{dV}{V}\right)} = P. \tag{4.67}$$

Hence from equation (4.66)

$$v = \sqrt{\frac{P}{\rho}} = \sqrt{\frac{nmc^2}{3nm}} = \left(\frac{\overline{c^2}}{3}\right)^{\frac{1}{2}}, \tag{4.68}$$

since the density $\rho = nm$.

Therefore

$$v = (\overline{u^2})^{\frac{1}{2}} \tag{4.69}$$

where $\overline{u^2}$ is the mean square molecular velocity in one direction. The value of $(\overline{u^2})^{\frac{1}{2}}$ is very nearly equal to u; this implies that the velocity of a sound wave is approximately equal to the mean molecular velocity in the direction of propagation. This at once tells us that the molecule is the messenger which carries the impulse through the gas. In this way we have linked the equation which describes wave propagation in terms of bulk properties [equation (4.66)], with a molecular process.

For an adiabatic wave the gas equation is

$$PV^\gamma = \text{constant}.$$

This leads to a value of the modulus

$$-\frac{dP}{\left(\dfrac{dV}{V}\right)} = \gamma P$$

and

$$v = \sqrt{\frac{\gamma P}{\rho}}. \tag{4.70}$$

Consequently the adiabatic wave velocity is $\sqrt{\gamma}$ times that given in equation (4.68). This increases v by about 20 per cent for sound in air, but does not change the basic conclusion that sound waves are propagated by molecules communicating an impulse to neighbouring molecules by collision with them.

4·6·2   *Why sound waves are adiabatic*

It has long been a tradition in text books of sound to assert that in practice sound waves are adiabatic, and not isothermal, because their frequencies are so high. There is a temperature rise at the compression and a cooling at the rarefaction, and this implies that the time taken for the wave to pass along (so reversing the position of compressions and rarefaction) is not sufficient for thermal equilibrium to be reached. Consequently the wave has not the time to become isothermal. This is true but we shall now show that this is because the *frequency is relatively low* and that high frequencies (in theory) would favour *isothermal* conditions. Basically this is because the time needed for temperature equilibriation to occur turns out to be proportional to the square of the wavelength, whereas the time available for this to occur is simply proportional to the wavelength. Thus equilibriation occurs only if the wavelength l is less than some critical length l_c as indicated schematically in figure 17(*a*).

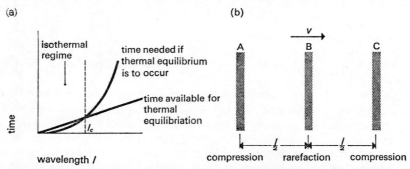

Figure 17.   (*a*) Time factors involved in achieving isothermal conditions in a sound wave; these are attainable only if the wavelength is less than l_c. (*b*) The compressions and rarefactions in a sound wave.

The following simple analysis shows how this arises:

Consider an instantaneous snapshot of a plane wave travelling from left to right with velocity v, frequency v and wavelength l. There are compressions at A, C, and a rarefaction at B [see figure 17(*b*)]. Suppose the heating at A and the cooling at B (under perfect adiabatic conditions) give a temperature difference between A and B of T. The temperature gradient is of course sinusoidal but for simplicity we shall assume that it is linear. Then we may write

$$\frac{dT}{dx} = -\frac{T}{\dfrac{l}{2}} = -\frac{2T}{l}.$$ (4.71)

As a result heat flows from A to B tending to equalize the temperature and so reducing the temperature gradient. As we are only interested in an order of magnitude calculation we shall ignore such changes and assume that the gradient remains at $-\dfrac{2T}{l}$.

If K is the thermal conductivity of the gas the rate of heat flow ΔQ per sq. cm. of section in time Δt is

$$\frac{\Delta Q}{\Delta t} = -K\frac{dT}{dx} = K\frac{2T}{l}. \tag{4.72}$$

Consequently the time Δt taken for a quantity of heat ΔQ to flow is

$$\Delta t = \frac{\Delta Q l}{2KT}. \tag{4.73}$$

Equalization of temperature between A and B demands that the heat ΔQ raises the temperature of the mass of gas between A and B by T, that is, we need an amount of heat,

$$\Delta Q = \text{volume} \times \text{number of molecules per c.c.} \times \text{specific heat per molecule} \times T$$

$$= \frac{l}{2}nc_P T. \tag{4.74}$$

(For an instantaneous snapshot of the wave, c_P is the appropriate specific heat.) Substituting this in equation (4.73) we obtain

$$\Delta t = \frac{l^2 nc_P}{4K}. \tag{4.75}$$

Now the time available for this equilibrium to occur cannot be greater than the time taken for the compression and rarefaction to change places, i.e. about $\tau/2$ where τ is the period of the wave. A more realistic estimate is $\dfrac{\tau}{4} = \dfrac{l}{4v}$. Consequently temperature equilibrium will occur, i.e. we shall approximate to isothermal conditions if

$$\frac{\tau}{4} > \Delta t \quad \text{i.e.} \quad \frac{l}{4v} > \frac{l^2 nc_P}{4K}.$$

This implies

$$l < \frac{K}{nc_P v} \quad \text{or} \quad v > \frac{nc_P v^2}{K}. \tag{4.76}$$

In a previous section we showed [see equation (4.60)] that for a perfect gas

$$K = \frac{n\bar{c}\lambda}{3}(c_V),$$

where λ is the mean free path. Hence isothermal conditions will be achieved if the frequency

$$\nu > \frac{3nc_p v^2}{n\bar{c}\lambda c_V}, \quad \text{i.e.} \quad \nu > \frac{3v^2}{\bar{c}\lambda} \tag{4.77}$$

ignoring the difference between c_P and c_V.

Since v is of the order of $\bar{c} \simeq 3 \times 10^4$ cm. sec.$^{-1}$ and λ is about 1000 Å. (10^{-5} cm.), this requires a frequency greater than 10^9 cycles sec.$^{-1}$. This is extremely high. Further if we do the same substitution for l we find that l must be less than $\dfrac{\bar{c}}{3v}\lambda$, that is, it must be less than the mean free path in the gas. Under these conditions a 'normal' sound wave cannot be propagated. We conclude that sound waves are adiabatic because their frequency is not high enough for them to become isothermal. In theory they could become isothermal at frequencies of the order of 10^{10} sec.$^{-1}$ but such frequencies could not be transmitted through a gas as a normal sound wave.

Chapter 5
Further theory of perfect gases

In this chapter we shall first describe a rather more sophisticated kinetic theory of perfect gases. Part of the exercise is purely computational and, although it looks more impressive, adds little to our physical understanding. There are, however, a number of points which emerge which are interesting and useful and which shed new light on some of the assumptions made in the simpler forms of the kinetic theory. We shall also discuss the velocity distribution in a gas and the thermal energy of its molecules.

5·1   A better kinetic theory

5·1·1   *Assumptions*

First we recapitulate our basic assumptions. Let us assume that we are dealing with a very large number of molecules uniformly distributed in density; that they have complete randomness of direction and velocity; that the collisions are perfectly elastic; that there are no intermolecular forces; and finally that the molecules have zero volume.

We now consider a way of describing their distribution in space. Thus to each molecule we attach a vector representing its velocity in magnitude and direction [figure 18(*a*)]. We then transfer these vectors (*not* the molecules) to

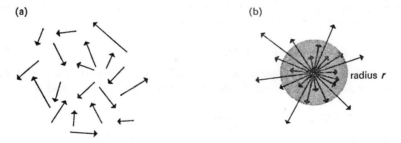

Figure 18.   (*a*) Velocity vectors are attached to the individual molecules. (*b*) The velocity vectors are all transferred to a common origin. A sphere of arbitrary radius *r* is drawn about the origin.

a common origin [figure 18(*b*)] and construct a sphere of arbitrary radius *r*, allowing the vectors to cut the sphere (if necessary by extending their length).

Then the velocity vectors intersect the sphere in as many points as there are molecules.

If we postulate randomness of molecular motion all directions are equally probable, so that these points will be uniformly distributed over the surface of the sphere. Suppose we consider ΔN molecules, where N is the total number present. These could be the total number of molecules in the vessel or the number per c.c. or the number per c.c. with a specified velocity range. Then the number of vector points corresponding to an element of area ΔA on the sphere will be

$$\Delta N = \frac{\Delta A}{4\pi r^2} N. \tag{5.1}$$

We can specify the element ΔA in terms of spherical co-ordinates θ and ϕ as shown in figure 19.

(a)   (b)

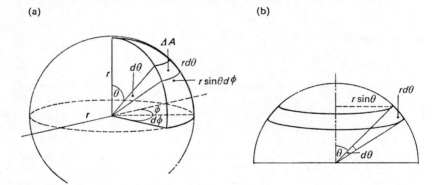

Figure 19. Specification of an element of area on the velocity sphere (a) in terms of angles θ and ϕ, (b) in terms only of angle θ.

$$\Delta A = rd\theta \times r \sin \theta \, d\phi. \tag{5.2}$$

Thus the number of molecules travelling in a direction between θ and $\theta + d\theta$ and between ϕ and $\phi + d\phi$ (relative to some arbitrary axes) is

$$\Delta N_{\theta, \phi} = \frac{\Delta A}{4\pi r^2} N = \frac{N}{4\pi} \times \sin \theta \, d\theta \, d\phi. \tag{5.3}$$

We see that r disappears from the result: ΔN depends only on the specified angles.

We could continue to use equation (5.3) in our subsequent calculations, but this would make the arithmetic more complicated than is necessary. In all the cases we are interested in, we only need to consider molecules travelling in a direction between θ and $\theta + d\theta$ irrespective of their ϕ position. That is to say it is sufficient for our purpose to make ΔA an annulus on the sphere lying between angles θ and $\theta + d\theta$ [figure 19(b)].

Then $\quad \Delta A = 2\pi r \sin \theta \, r d\theta,$ (5.4)

so that the total number of molecules travelling in this direction is

$$\Delta N_\theta = \frac{\Delta A}{4\pi r^2} N = \tfrac{1}{2} \sin \theta \, d\theta N.$$ (5.5)

5·1·2 *Number of collisions with a solid wall*

We use the above result to calculate the number of molecules colliding per second with a square centimetre of a solid surface. Let us first divide the molecules into groups according to their velocities. Suppose there are n_c molecules per c.c. with velocity between c and $c + dc$. Then the number out of this group that at any instant are travelling towards the surface from a direction θ, $\theta + d\theta$ is

$$\Delta n_{c,\theta} = \tfrac{1}{2} \sin \theta \, d\theta \times n_c.$$ (5.6)

The only molecules that hit the surface of area α in time dt are those contained within a prism of basal area α and of sloping height cdt [see figure 20(a)]. The

(a)

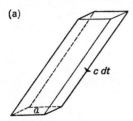

Figure 20. (a) The number of molecules hitting the area α in time dt is equal to the number of molecules within the prism of basal area α and sloping height cdt.

volume of this is $cdt \, \alpha \cos \theta$. Hence the number of molecules in this group hitting α in time dt is

$$\Delta n_{c,\theta} \times cdt \, \alpha \cos \theta.$$ (5.7)

The number striking each sq. cm. per sec. is this number divided by αdt

$$= \tfrac{1}{2} \sin \theta \cos \theta \, d\theta \, n_c c.$$ (5.8)

For all values of θ from 0 to $\pi/2$, i.e., for the whole of the 'half-space' above the surface the total collision rate is

$$\int_0^{\frac{\pi}{2}} \tfrac{1}{2} \sin \theta \cos \theta \, d\theta \, n_c c = \tfrac{1}{4} n_c c.$$ (5.9)

67   Further theory of perfect gases

For all molecules of all velocity groups the number striking each sq. cm. per sec. is then

$$\tfrac{1}{4} \Sigma n_c c = \tfrac{1}{4} n \bar{c}, \tag{5.10}$$

where n is the total number per c.c. and $\bar{c}$ is the mean velocity defined by $\bar{c} = \dfrac{\Sigma n_c c}{n}$. This result differs from that given in the previous chapter [equation (4.52)] where the fraction in front of $n\bar{c}$ was found to be $\tfrac{1}{6}$ instead of $\tfrac{1}{4}$.

5·1·3 Gas equation of state

We now consider the momentum transfer per second per sq. cm. of wall, i.e., the pressure exerted by the gas. We first consider the group of molecules n_c per c.c. whose velocity is between c and $c + dc$. Of these consider the group approaching the surface at a direction θ, $\theta + d\theta$; each molecule brings with it momentum mc. The vertical component is, therefore, $mc \cos\theta$ and after collision it is reversed so that the net momentum transferred to the wall is $2mc \cos\theta$. The horizontal component $mc \sin\theta$ is unchanged so that no tangential momentum is communicated to the wall. The vertical momentum change per second transferred to each sq. cm. of wall, i.e. the pressure, is

$$2mc \cos\theta \times \text{(number hitting each sq. cm. per sec.)}$$
$$= 2mc \cos\theta \, \tfrac{1}{2} \sin\theta \cos\theta \, d\theta \, n_c c$$
$$= mn_c c^2 \cos^2\theta \sin\theta \, d\theta. \tag{5.11}$$

Integrating over the whole half-space we obtain the pressure produced by this group of molecules:

$$mn_c c^2 \int_0^{\frac{\pi}{2}} \cos^2\theta \sin\theta \, d\theta$$
$$= mn_c c^2 \left[-\frac{\cos^3\theta}{3} \right]_0^{\frac{\pi}{2}} = \tfrac{1}{3} mn_c c^2. \tag{5.12}$$

For all groups of molecules we thus obtain

$$p = \tfrac{1}{3} mn\overline{c^2}, \tag{5.13}$$

where the mean square velocity $\overline{c^2}$ is given by $\dfrac{\Sigma n_c c^2}{n}$.

5·1·4 Transport phenomena

Consider for example the viscosity of a gas in terms of the momentum transport across a plane as described in the previous chapter. The flow is assumed streamlined parallel to the XX plane, the velocity gradient du/dy in the steady state is constant. We have to calculate the transport of horizontal momentum across each sq. cm. in the plane XX per sec. Consider a small area element $d\alpha$ in XX. The only molecules we are interested in are those which on an average made their last collision at a distance λ from the element. Draw a hemisphere of radius λ with the element as centre [see figure 20(b)]. Those

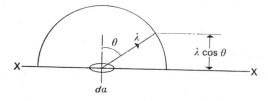

Figure 20. (b) Transport of momentum through element of area $d\alpha$. A hemisphere of radius λ is constructed with $d\alpha$ as centre. All molecules from this hemisphere have, on average, just made their last collision before proceeding towards $d\alpha$, from half of the gas.

coming from the direction θ, $\theta + d\theta$, are at a distance $\lambda \cos \theta$ from XX so that these molecules bring additional horizontal momentum $m\lambda \cos \theta \left(\dfrac{du}{dy}\right)$.

The number coming from this direction and striking each sq. cm. of the element per sec. is given by equation (5.8). The transfer of horizontal momentum per sq. cm. per sec. is therefore:

$$m\lambda \cos \theta \left(\frac{du}{dy}\right) \times \tfrac{1}{2} \sin \theta \cos \theta \; d\theta \; n_c c. \tag{5.14}$$

If we integrate this over the whole of the top half-space we obtain

$$\tfrac{1}{6} m\lambda n_c c \times \left(\frac{du}{dy}\right). \tag{5.15}$$

There is a similar contribution from the half-space below so that the tangential stress in the plane XX is simply

$$\tfrac{1}{3} m\lambda n_c c \left(\frac{du}{dy}\right). \tag{5.16}$$

If we equate this to the viscous stress $\eta \dfrac{du}{dy}$ we obtain

$$\eta = \tfrac{1}{3} m n_c c \lambda, \tag{5.17}$$

for molecules of this group. For all groups of molecules

$$\eta = \tfrac{1}{3} m n \bar{c} \lambda. \tag{5.18}$$

69 Further theory of perfect gases

The result is identical with that obtained from the Joule classification.

We may explain this result in a simple way. The number of molecules striking each sq. cm. in XX per sec. is $\frac{1}{4}n\bar{c}$ instead of $\frac{1}{6}n\bar{c}$ as in the Joule derivation, but whereas in the Joule derivation each molecule is assumed to come from a constant distance λ from plane XX, in the present treatment the last collision occurs on the hemisphere of radius λ, so that the last collision distance measured normal to XX ranges from 0 to λ. The average effective distance is $\frac{2}{3}\lambda$. The horizontal momentum transfer is thus the same in both models.

5·1·5 Mean free path λ and collision frequency v

We now extend our treatment to allow for the finite diameter σ of the molecules. As we saw in the previous chapter, a very simple result for the average distance λ between each collision is given by

$$\lambda = \frac{1}{\pi\sigma^2 n}. \tag{5.19}$$

There are many sophisticated ways of improving this relation but they only change λ by a small numerical factor.

We now define the collision frequency v as the average number of collisions per second made by each molecule.

Then

$$v = \frac{\text{Average distance travelled per sec.}}{\text{Average distance between collisions}}$$

$$v = \frac{\bar{c}}{\lambda}. \tag{5.20}$$

This result does not depend on the way in which λ is derived.

5·1·6 Distribution of free paths: survival equation

We wish to follow the fortunes of a group of N molecules. As they travel they collide with themselves and other molecules. Can we estimate the number that, at any specified stage, have not yet made a collision?

Suppose that, at some instant, n have survived without collision. If each is allowed to travel a further distance dx along its free path further collisions may occur. We *assume* that this number of collisions is proportional to both n and dx. Hence the number removed by these collisions is proportional to ndx. To show that n is decreased we write:

$$dn = -Pndx, \tag{5.21}$$

where P is the collision probability for the gas. We assume this to be a constant for a given gas under specified conditions. Hence,

$$\frac{dn}{n} = -Pdx$$

$$n = N\,e^{-Px} \tag{5.22}$$

and $\qquad dn = -PN\,e^{-Px}dx.$ $\hfill$ (5.23)

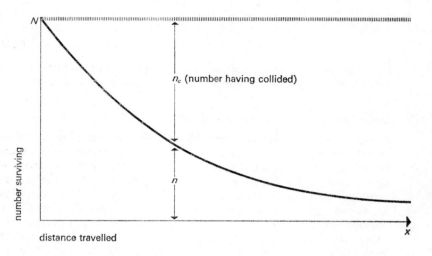

Figure 21. Graph showing survival of an initial group of N molecules. After travelling a distance of x, n have survived without collision whilst the number n_c suffering collision is given by $n_c = N-n$.

Since n is the number surviving after travelling a distance x, the number n_c which have suffered collision (see figure 21) is simply

$$n_c = N-n.$$

Hence

$$dn_c = d(N-n) = -dn,$$

or

$$dn_c = +PN\,e^{-Px}dx.$$

This is the number of molecules with mean free path between x and $x+dx$. The average mean free path x is given by

$$\lambda = \frac{\displaystyle\int_{n=0}^{n=N} x\,dn_c}{\displaystyle\int_{0}^{N} dn_c}$$

71   Further theory of perfect gases

i.e., $\lambda = \dfrac{\displaystyle\int_{x=0}^{x=\infty} xPN\,e^{-Px}dx}{N}$

$$= -\left[x\,e^{-Px} + \frac{e^{-Px}}{P}\right]_0^\infty = \frac{1}{P}.$$

Consequently, equation (5.22) becomes

$$n = N e^{-\frac{x}{\lambda}}. \tag{5.24}$$

This is the basic 'survival equation'. n is the number of molecules out of a group N that have not yet made a collision after travelling a distance x.

5·1·7   Collisions with surfaces: a final treatment

We may now ask a simple but important question in relation to the kinetic theory as we have so far developed it. We have always assumed that if somewhere there is a group of molecules travelling in some specified direction we can calculate the number striking a sq. cm. of a surface per second by completely ignoring intermolecular collisions. The previous section shows that this is quite unjustified. Molecules a long way from the surface never get there: they are diverted by collisions on the way with themselves and other molecules.

Consider at any instant an element of volume dV in a gas of uniform density containing n molecules per c.c. [figure 22(a)]. The number in the

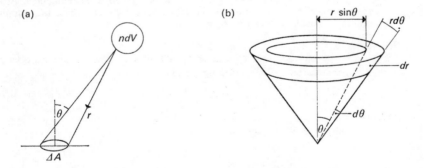

(a)   (b)

Figure 22.   (a) Molecules in element of volume dV, after their last collision, start travelling towards ΔA. (b) Conical annulus of volume $2\pi r \sin\theta\, rd\theta\, dr$ defines dV.

element is ndV.

If v is the collision frequency, the number of collisions occurring within dV in time dt is

$$\tfrac{1}{2}v(ndV)dt. \tag{5.25}$$

The $\frac{1}{2}$ is introduced since otherwise each collision would be counted twice. Each collision starts off two new free paths so the total number of free paths originating in time dt from the element dV is

$$\nu n dV \, dt.$$

Because all directions of molecular motion are equally probable in any element of the gas, these molecules start off uniformly in all directions. The fraction heading towards the area ΔA is $\dfrac{\Delta\omega}{4\pi}$, where $\Delta\omega$ is the solid angle subtended by ΔA at the volume element.

Clearly,

$$\Delta\omega = \frac{\Delta A \cos\theta}{r^2}.$$

The number heading towards ΔA is thus

$$\Delta n = \frac{\Delta\omega}{4\pi} \times \nu n dV \times dt. \tag{5.27}$$

However, from the survival equation, the number of molecules reaching ΔA without having made a collision, and so having been eliminated is

$$\Delta n_1 = \Delta n \, e^{-\frac{r}{\lambda}}.$$

Consequently, the number of molecules leaving dV in time dt and ultimately reaching ΔA without having made an intermediate collision is

$$\Delta n_1 = \frac{1}{4\pi} \frac{\Delta A \cos\theta}{r^2} \nu n \, e^{-\frac{r}{\lambda}} \, dV \, dt. \tag{5.28}$$

We note that Δn_1 depends on the time element dt, not on the absolute time. This is because on the average any given element of volume appears unchanged. (To some extent this observation, in itself, implies the conclusion we shall draw in the following paragraphs.) We may thus choose our time element dt for each volume element dV at various times, but in such a way that all the undeflected molecules from all the volume elements arrive at ΔA during the same time interval dt. This enables us to integrate for the whole volume of gas above ΔA. Once again we choose as our element of volume a conical annulus of radius $r \sin\theta$, thickness $rd\theta$, height dr [figure 22(b)], so that

$$dV = 2\pi r \sin\theta \, rd\theta \, dr. \tag{5.29}$$

We substitute in equation (5.28) and integrate for $r = 0$ to $r = \infty$, and for $\theta = 0$ to $\theta = \dfrac{\pi}{2}$. This gives the total number of molecules, from all half-

73   Further theory of perfect gases

space, from all directions striking ΔA in time dt as:

$$vn\,\Delta A\,dt \int_0^\infty \int_0^{\frac{\pi}{2}} \frac{\cos\theta\sin\theta}{2}\,d\theta\,e^{-\frac{r}{\lambda}}\,dr \qquad (5.30)$$

$$= \tfrac{1}{4}vn\lambda\,\Delta A\,dt.$$

The number striking per sq. cm. per sec. is then

$$\tfrac{1}{4}vn\lambda. \qquad (5.31)$$

But, irrespective of our model for calculating λ,

$$v = \frac{\bar{c}}{\lambda}.$$

Hence the number striking each sq. cm. per sec. is

$$\tfrac{1}{4}n\bar{c}. \qquad (5.32)$$

This is the same as the result we have already obtained. It implies that although molecules are continuously colliding and changing their directions and velocities, other molecules are replacing the ones that are eliminated. For this reason the simple picture which ignores intermolecular collisions in calculating wall collisions, gas pressure and transport phenomena yields the correct answer.

5·2 Sedimentation

5·2·1 *Sedimentation under gravity*

Consider a gas column of unit cross-section and at uniform temperature T (figure 23). We shall show that because of gravity there is a density (and

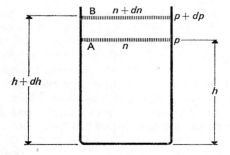

Figure 23.   Decrease of density and pressure with height of an ideal gas.

pressure) gradient in the gas. Our increments are measured positively in the direction of increasing height h.

At the level A, height h, let the pressure be p and the molecular density n per c.c. At level B, height $h + dh$, the pressure is $p + dp$ and the density $n + dn$.

The resultant force on the layer of gas between A and B is a pressure dp downwards and a gravitational force $n \times dh \times mg$ downwards. For equilibrium

$$n\, dh\, mg + dp = 0. \tag{5.33}$$

Since $p = \tfrac{1}{3}nm\overline{c^2}$ and $\overline{c^2}$ is constant at all levels (because of constant temperature)

$$dp = \tfrac{1}{3}m\overline{c^2}dn.$$

Equation (5.33) becomes

$$\frac{dn}{n} = \frac{-mg}{\tfrac{1}{3}m\overline{c^2}}\, dh.$$

Integrating from $h = 0$ to h and from n_0 to n we obtain

$$\ln\frac{n}{n_0} = \frac{-mg}{\tfrac{1}{3}m\overline{c^2}}\, h = \frac{-Nmg}{\tfrac{1}{3}Nm\overline{c^2}}\, h,$$

or $\qquad \ln\dfrac{n}{n_0} = \ln\dfrac{p}{p_0} = -\dfrac{Mg}{RT}\, h. \tag{5.34}$

For air, substituting $M \simeq 30$, $R = 8 \cdot 4 \times 10^7$, $T = 300$, we find that at a height of 1000 ft. the pressure is reduced by about one thirtieth of an atmosphere.

5·2·2   *Sedimentation of particles*

If particles of mass m and density ρ are suspended in a liquid of density ρ_0, then if there is complete thermal equilibrium, the concentration of particles is given by

$$\ln\frac{n}{n_0} = -\frac{Nmgh}{RT}\frac{(\rho - \rho_0)}{\rho}. \tag{5.35}$$

An examination of the variation of n with height for a fine suspension was used by J. Perrin in 1908 to calculate N_0. He obtained a value of 6×10^{23}, which is virtually identical with the value of Avogadro's number obtained by other methods. This shows that fine particles in thermal equilibrium behave like a gas of molecular mass m. It is quite wrong to imagine that this is simply because they are buffeted around by the molecules of the liquid in which they are suspended. They possess this property simply by virtue of being in thermal equilibrium with themselves and their surroundings. Of course the collisions between the particles and the liquid molecules ensure that the whole system is in thermal equilibrium, but the kinetic theory applies to the particles themselves just as it does to the molecules of a conventional gas. Equation

(5.35) also shows that by increasing g the major fraction of the particles will be formed in a thin layer near the bottom of the vessel. This is the basis of the action of a centrifuge.

5·2·3   The Boltzmann distribution

We may rewrite equation (5.34) in the form

$$\ln \frac{n}{n_0} = -\frac{Mg}{RT}h = -\frac{mg}{kT}h$$

where $k = \dfrac{R}{N} =$ the Boltzmann constant.

Then $n = n_0 \exp\left(-\frac{mg}{kT}h\right)$.   (5.36)

Since the temperature is constant throughout the gas the kinetic energy is constant at all levels; only the potential energy has been changed, by amount mgh. Then mgh is the amount by which the energy of a molecule at h exceeds that at ground level; call this ε. Then

$$n = n_0 \exp\left(-\frac{\varepsilon}{kT}\right) .$$ (5.37)

Boltzmann showed that for any equilibrium distribution of molecules this relation generally holds. We may write

$$\begin{pmatrix} \text{number of molecules at energy} \\ \text{level } \varepsilon \text{ above ground level} \end{pmatrix} = \begin{pmatrix} \text{number of molecules} \\ \text{at ground level} \end{pmatrix} \times \exp\left(-\frac{\varepsilon}{kT}\right).$$

Again for two energy states ε_1 and ε_2

$$\frac{n_1}{n_2} = \frac{\exp\left(-\dfrac{\varepsilon_1}{kT}\right)}{\exp\left(-\dfrac{\varepsilon_2}{kT}\right)} = \exp\left(-\frac{\Delta\varepsilon}{kT}\right) ,$$ (5.38)

where $\Delta\varepsilon$ is the energy gap between states 1 and 2. The Boltzmann function thus describes the relative population of energy states in an equilibrium system.

5·3   Temperature variation of reaction rates

A chemical reaction can only occur spontaneously if the final system has a lower free energy state than the initial system. In general however, before the reaction can take place one or more of the reactants must be excited to a higher energy state (see figure 24). The rate of forward reaction is then

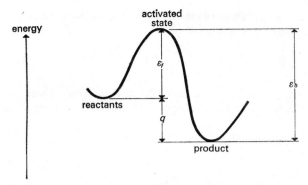

Figure 24. Forward reaction has an activation energy ε_f, the backward reaction an activation energy ε_b. The heat of reaction is $q = \varepsilon_b - \varepsilon_f$.

determined by the number of molecules which have enough energy to get over this barrier. The forward reaction rate is proportional to $\exp\left(-\dfrac{\varepsilon_f}{kT}\right)$ where ε_f is the activation energy.

It is interesting to note that the back reaction rate will be proportional to $\exp\left(-\dfrac{\varepsilon_b}{kT}\right)$, and that the heat of reaction is equal to the difference between the activation energies of forward and backward reaction,

$$q = \varepsilon_b - \varepsilon_f.$$

5·4 Distribution of velocities in a perfect gas

5·4·1 *Velocity distribution in a one-, two- and three-dimensional gas*

We expect to find a wide distribution of molecular velocities in a gas, because even if we could start off with all the molecules travelling in different directions with equal velocities, random collisions are bound to speed some and retard others. Consequently we expect to find the velocities ranging from 0 to ∞* but grouped around a definite average value determined by the temperature. The velocity distribution is, in fact, of the same type as the Boltzmann function.

If there are no intermolecular forces the only energy the individual molecule has is its kinetic energy $\frac{1}{2}mc^2$. (We ignore here vibrational or rotational energies which are assumed to be unchanged by the gas-collision process.)

* Because of relativity effects an infinite velocity would imply an infinite effective mass. However, the fraction of molecules with velocities high enough to produce appreciable relativity effects is negligible. In what follows we shall assume that relativity effects may be ignored and that the velocity distribution extends to infinity although, of course, infinite velocities cannot in fact occur.

77 Further theory of perfect gases

Then according to the Boltzmann distribution the population density of molecules with velocity c is proportional to

$$\exp\left(-\frac{mc^2}{2kT}\right).$$
(5.39)

If we attach velocity vectors to the molecules and transpose them to a single origin [see figure 25(a)], the number of velocity vectors ending in the

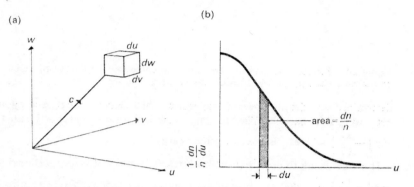

Figure 25. Velocity vector diagrams (a) for molecules with velocities between c and $c+du+dv+dw$, (b) velocity distribution for a one-dimensional gas. The area of the shaded band gives the fraction of molecules with velocities between u and $u+du$ in the u direction.

element du, dv, dw gives us the number of molecules with velocities between u and $u+du$, v and $v+dv$, w and $w+dw$. This number is then

$$dn = A \exp\left(-\frac{mc^2}{2kT}\right) du\, dv\, dw.$$
(5.40)

Before proceeding to use this relation we may split up c^2 into its three components $u^2 + v^2 + w^2$ so that

$$dn = A \exp\left(-\frac{mu^2}{2kT}\right) \exp\left(-\frac{mv^2}{2kT}\right) \exp\left(-\frac{mw^2}{2kT}\right) du\, dv\, dw.$$
(5.41)

Since each of the three components must have the same velocity distributions, the quantity

$$A^{\frac{1}{3}} \exp\left(-\frac{mu^2}{2kT}\right) du$$

gives the number of molecules with velocity between u and $u+du$ irrespective of the v and w components. The fraction of molecules in this category is

$$\frac{dn}{n} = B \exp\left(-\frac{mu^2}{2kT}\right) du. \tag{5.42}$$

The constant B may be determined by observing that, if this relation is integrated for all possible velocities from $-\infty$ to $+\infty$, the resulting fraction must be unity. Then

$$\int_{-\infty}^{\infty} B \exp\left(-\frac{mu^2}{2kT}\right) du = 1. \tag{5.43}$$

Putting $\dfrac{m}{2kT} = a$ \hfill (5.44)

$$B \int_{-\infty}^{\infty} e^{-au^2} du = 1.$$

From the table of integrals at the end of the chapter

$$\int_{-\infty}^{\infty} e^{-au^2} du = 2 \int_{0}^{\infty} e^{-au^2} du = \left(\frac{\pi}{a}\right)^{\frac{1}{2}} = \left(\frac{2\pi kT}{m}\right)^{\frac{1}{2}}.$$

Hence $\quad B = \left(\dfrac{m}{2\pi kT}\right)^{\frac{1}{2}}.$

Consequently

$$\frac{dn}{n} = \left(\frac{m}{2\pi kT}\right)^{\frac{1}{2}} \exp\left(-\frac{mu^2}{2kT}\right) du. \tag{5.45}$$

How can we plot this function? We bring the term du over to the left-hand side so that we have

$$\frac{dn}{n}\frac{1}{du} \quad \text{or} \quad \frac{1}{n}\frac{dn}{du} = \left(\frac{m}{2\pi kT}\right)^{\frac{1}{2}} \exp\left(-\frac{mu^2}{2kT}\right). \tag{5.46}$$

We may calculate the right-hand side and plot it as a function of u [see figure 25(b)]. If we take a strip of width du its area is then dn/n which is the fraction of molecules with velocity between u and $u+du$. We see that the greatest fraction occurs for zero velocity; this is because of the exponential nature of the Boltzmann factor, and the fact that the 'one-dimensional' gas directly reflects this distribution.

The average energy in the u direction is given by

$$\bar{\varepsilon} = \left[\int_{-\infty}^{\infty} \left(\frac{mu^2}{2}\right) dn\right]\frac{1}{n} \tag{5.47}$$

$$\bar{\varepsilon} = \frac{m}{2}\left(\frac{m}{2\pi kT}\right)^{\frac{1}{2}} \int_{-\infty}^{\infty} u^2 e^{-au^2} du = \frac{m}{2}\left(\frac{a}{\pi}\right)^{\frac{1}{2}} \int_{-\infty}^{\infty} u^2 e^{-au^2} du.$$

79 Further theory of perfect gases

From the table of integrals at the end of the chapter

$$\int_{-\infty}^{\infty} u^2 e^{-au^2} \, du = \frac{1}{2} \left(\frac{\pi}{a^3} \right)^{\frac{1}{2}}.$$

Hence $\quad \bar{\varepsilon} = \dfrac{m}{2} \left(\dfrac{a}{\pi} \right)^{\frac{1}{2}} \dfrac{1}{2} \left(\dfrac{\pi}{a^3} \right)^{\frac{1}{2}} = \dfrac{m}{4} \times \dfrac{1}{a} = \dfrac{m}{4} \dfrac{2kT}{m},$

or $\qquad \bar{\varepsilon} = \dfrac{kT}{2}.$ $\hfill$ (5.48)

Similarly the fraction of molecules with velocity components between u and $u+du$ and between v and $v+dv$ is simply the product of the two individual probabilities.

$$\frac{dn}{n} = \left(\frac{m}{2\pi kT} \right) \exp \left[-\frac{m(u^2 + v^2)}{2kT} \right] du \, dv. \tag{5.49}$$

Finally the fraction of molecules with velocity components between u and $u+du$, v and $v+dv$, w and $w+dw$ is

$$\frac{dn}{n} = \left(\frac{m}{2\pi kT} \right)^{\frac{3}{2}} \exp \left[-\frac{m(u^2 + v^2 + w^2)}{2kT} \right] du \, dv \, dw. \tag{5.50}$$

Generally we are not interested in the dependence of n on the individual components of velocity. For practically all purposes we wish rather to know the number of molecules with velocities between c and $c+dc$, irrespective of direction. We may calculate this easily by drawing a sphere of radius c and another of radius $c+dc$ [see figure 26(a)]. The number of velocity vectors ending in the spherical shell is then the number required. In velocity space the volume of the shell is $4\pi c^2 dc$ and this replaces $du \, dv \, dw$ in equation (5.50). We then obtain

$$\frac{1}{n} \frac{dn}{dc} = 4\pi c^2 \left(\frac{m}{2\pi kT} \right)^{\frac{3}{2}} \exp \left(-\frac{mc^2}{2kT} \right). \tag{5.51}$$

This is plotted in figure 26(b). We note that in any integration to give total quantities, c ranges from zero to ∞, not from $-\infty$ to $+\infty$.

[The reader, if he is interested only in the three-dimensional gas, may pass at once from equation (5.40) to equation (5.51) by writing equation (5.40) in the form

$$\frac{dn}{n} = C \exp \left(-\frac{mc^2}{2kT} \right) 4\pi c^2 dc.$$

By specifying that $\displaystyle\int_{c=0}^{c=\infty} \frac{dn}{n} = 1$, he will find that C has the value $\left(\dfrac{m}{2\pi kT} \right)^{\frac{3}{2}}$.
This results in equation (5.51).]

Rewriting equation (5.51) in the form

$$\frac{dn}{n} = 4\pi \left(\frac{a}{\pi}\right)^{\frac{3}{2}} e^{-ac^2} c^2 dc, \qquad (5.52)$$

and using the table of integrals at the end of the chapter we may readily calculate the following three representative velocities: the most probable velocity c_m, that is the velocity for which equation (5.52) is a maximum, the average velocity $\bar{c}$, and the root mean velocity $(\overline{c^2})^{\frac{1}{2}}$. We have

most probable
$$c_m = \sqrt{2}\left(\frac{kT}{m}\right)^{\frac{1}{2}}; \qquad (5.53a)$$

average
$$\bar{c} = \int_0^\infty c\,\frac{dn}{n} = \sqrt{\frac{8}{\pi}\left(\frac{kT}{m}\right)^{\frac{1}{2}}}, \qquad (5.53b)$$

r.m.s. $(\overline{c^2})^{\frac{1}{2}} = \left[\int_0^\infty \frac{c^2 dn}{n}\right]^{\frac{1}{2}} = \sqrt{3}\left(\frac{kT}{m}\right)^{\frac{1}{2}}.$ $\qquad (5.53c)$

Before leaving this we note one of the confusing features of probability distributions. According to equation (5.39) the population density is a maximum when $c = 0$. This is also true, as we saw above, of the velocity distribution in a 'one-dimensional' gas. As soon, however, as we specify velocity elements in two or three dimensions the most probable velocity (as distinct from the probability density) no longer corresponds to zero velocity. This is specifically brought out in equation (5.51) and figure 26 for a

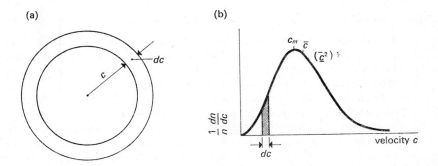

Figure 26.   (*a*) Velocity vector diagram for molecules with velocities between *c* and *c*+*dc*. This is given by the number of points ending the velocity vectors, which lie within the spherical shell of radii *c* and *c*+*dc*. (*b*) Maxwellian velocity distribution for a three-dimensional gas. The shaded band gives the fraction of molecules with velocities between *c* and *c*+*dc*. The approximate positions of the most probable velocity *cm*, the mean velocity *c̄* and the root mean square velocity $(\overline{c^2})^{\frac{1}{2}}$ are indicated.

81   Further theory of perfect gases

three-dimensional gas. A similar point will arise when we discuss the behaviour of rubber molecules in Chapter 8.

5·4·3 Number of molecules striking a surface per second

The number of molecules striking a square centimetre of surface per second may be directly obtained from equation (5.45). If the surface considered is at right angles to the u direction the v and w components are not involved. Consider molecules with velocity u, $u+du$; we construct a prism of unit cross-section and length u. Then all the molecules in this volume will reach the surface in one second. This number is $u\,dn$ where dn is the number per c.c. with velocity u in the sense *towards* the surface. The total number required is the integral of $u\,dn$ for all values of u from 0 to infinity: this choice of limits ensures that we are considering only those molecules moving towards the surface.

The number is

$$n \int_0^\infty \left(\frac{m}{2\pi kT}\right)^{\frac{1}{2}} \exp\left(-\frac{mu^2}{2kT}\right) u\,du \tag{5.45a}$$

$$= n \left(\frac{m}{2\pi kT}\right)^{\frac{1}{2}} \left(-\frac{kT}{m}\right) \exp\left[-\frac{mu^2}{2kT}\right]_0^\infty \tag{5.45b}$$

$$= n \left(\frac{kT}{2\pi m}\right)^{\frac{1}{2}} = \frac{n}{4}\left(\frac{8}{\pi}\times\frac{kT}{m}\right)^{\frac{1}{2}}. \tag{5.45c}$$

Comparison with equation (5.53b) shows that this is equal to $\frac{1}{4}n\bar{c}$.

5·4·4 Maxwell's derivation of the velocity distribution in a gas

The velocity distribution derived above makes use of the Boltzmann distribution. It is interesting to see how Maxwell derived the velocity distribution before the more general ideas of Boltzmann had been developed. The following is based on Maxwell's paper published in 1860.

Let n be the total number of molecules, and let u, v, w, be the components of the velocity of each molecule in three rectangular directions. Then if the number of molecules for which velocity component u lies between u and $u+du$ is $n\,f(u)\,du$, where $f(u)$ is the function to be determined, the number of molecules for which v lies between v and $v+dv$ will be $n\,f(v)\,dv$ and similarly for w, where f always stands for the same function since there is no preferred direction in the gas.

Now the existence of the velocity n does not in any way affect that of the velocities v and w since these are all at right angles to each other and independent. Consequently the number of molecules whose velocity components lie simultaneously between u and $u+du$, between v and $v+dv$ and between w and $w+dw$ is

$$dn = n f(u) f(v) f(w) \, du \, dv \, dw. \tag{5.54a}$$

Let this now refer to all those molecules which have a resultant velocity c where

$$c^2 = u^2 + v^2 + w^2. \tag{5.54b}$$

We now consider the condition under which the components u, v, w, can vary whilst c remains constant. This is found by differentiating equation (5.54b) and putting $dc = 0$; we obtain

$$u \, du + v \, dv + w \, dw = 0. \tag{5.54c}$$

Since no direction is preferred over any other, it follows that dn in equation (5.54a) must remain constant whatever the individual values of u, v and w, provided these satisfy equation (5.54c). This implies that

$$\frac{d}{du}(dn) \, du + \frac{d}{dv}(dn) \, dv + \frac{d}{dw}(dn) \, dw = 0. \tag{5.54d}$$

From equation (5.54a) this gives

$$f'(u) \, du \, f(v) f(w) + f'(v) \, dv \, f(u) f(w) + f'(w) \, dw \, f(u) f(v) = 0$$

where $f'(u)$ is the differential of $f(u)$ with respect to u, etc. Dividing throughout by $f(u) f(v) f(w)$ we obtain

$$\frac{f'(u)}{f(u)} \, du + \frac{f'(v)}{f(v)} \, dv + \frac{f'(w)}{f(w)} \, dw = 0. \tag{5.54e}$$

In order to solve equations (5.54c) and (5.54e), we multiply equation (5.54c) by an arbitrary constant λ and add it to equation (5.54e). The result is

$$\left[\frac{f'(u)}{f(u)} + \lambda u \right] du + \left[\frac{f'(v)}{f(v)} + \lambda v \right] dv + \left[\frac{f'(w)}{f(w)} + \lambda w \right] dw = 0.$$

Each term must identically be equal to zero, and since du, dv, dw, although very small are not themselves zero, the quantities in the brackets must be zero. Hence

$$\frac{f'(u)}{f(u)} + \lambda u = 0 \quad \text{or} \quad \frac{f'(u)}{f(u)} = -\lambda u.$$

Integrating this relation we have

$$\ln f(u) = -\frac{\lambda u^2}{2} + A$$

where A is an integration constant. This may be written in the form

$$f(u) = B \exp\left(-\frac{\lambda u^2}{2} \right).$$

83   Further theory of perfect gases

As Maxwell remarks, if λ were negative, the number of molecules would be infinite, so consequently, λ must be positive. The number of molecules with velocities between u and $u + du$ then becomes

$$dn = nf(u)\, du = nB \exp\left(-\frac{\lambda u^2}{2}\right).$$

There is no way of establishing the value of λ except by calculating the mean square velocity $\overline{c^2}$ and making use of the Gas equation: $\frac{1}{2}m\overline{c^2} = kT$. It is then found that λ has the value m/kT. The result is

$$dn = nB \exp\left(-\frac{mu^2}{2kT}\right) du,$$

which is identical with equation (5.42).

5·4·5 Experimental determination of velocity distribution

In most of the earlier experimental studies of the velocity distribution, a furnace was used to produce a vapour of metal atoms, which behave like a monatomic gas, at the temperature of the enclosure. They were allowed to condense on a cold surface and by the ingenious use of moving shutters atoms with different velocities were caught at different points on the surface. The intensity of condensed atoms provides then a measure of the relative number

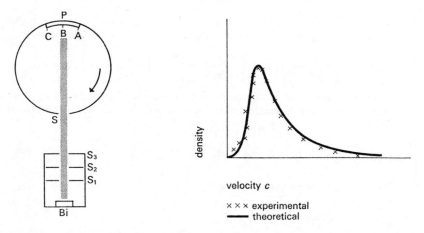

Figure 27. (a) Method due to Zartman and Ko for determining the velocity distribution of bismuth atoms; the atoms are evaporated, collimated by the slits S_1, S_2, S_3 and enter the rotating drum. They fall on the cold plate P where they condense. (b) Results obtained in a later experiment by Esterman, Simpson and Stern.

of atoms within that velocity range. One method due to I. F. Zartman (1931) and C. C. Ko (1934) is illustrated in figure 27(a). Atoms of bismuth are produced by a furnace and the vapour is collimated by a series of slits, S_1, S_2, S_3; the vapour beam reaches a drum rotating at a speed of 6000 rev. min.$^{-1}$ and can only enter the drum at the slit S. The atoms then strike the plate P where they condense; the fastest ones reaching A, and slower ones B and the slowest C.

A more elegant and refined experiment by I. Esterman, O. C. Simpson and O. Stern (1947) gave results shown in figure 27(b). The agreement with the Boltzmann distribution is surprisingly good.

5·5 Thermal energy of molecules

5·5·1 *Specific heats, number of degrees of freedom*

We have already seen that for a perfect gas in thermal equilibrium each molecule possesses an average thermal energy of translation of amount $\frac{3}{2}kT$. Since a molecule has 3 independent degrees of translational freedom we may associate energy $\frac{1}{2}kT$ with each degree of freedom. This may be generalized to apply to all particles in thermal equilibrium; however, for macroscopic bodies the resulting velocity is rather small. For example the mean thermal velocity of a cricket ball in equilibrium with its surroundings at room temperature (say 300°K.) is about 10^{-8} cm. sec.$^{-1}$.

The result may be generalized even further. If we define the number of degrees of freedom as the number of independent squared terms that enter into the total energy of a particle, then each has associated with it an average thermal energy of amount $\frac{1}{2}kT$. For example the kinetic energy due to translation is determined solely by (linear velocity)^2 in each of these independent co-ordinates; each degree of freedom has thermal energy $\frac{1}{2}kT$. The energy of rotation of a body about a specified axis is determined by (angular velocity)^2, so that each of these degrees of freedom has energy $\frac{1}{2}kT$. Vibrations present a special problem. If the equation of motion, for a body of mass m, is

$$\ddot{x} = -\omega^2 x, \text{ with solution } x = A \sin \omega t, \tag{5.55}$$

the kinetic energy at any instant is $\frac{1}{2}m\dot{x}^2$, whilst the potential energy is easily shown to be equal to $\frac{1}{2}m\omega^2 x^2$. By substituting in equation (5.55) we find that the sum of kinetic and potential energy is $\frac{1}{2}mA^2\omega^2$, so it is independent of x and $\dot{x}$. Nevertheless at any instant the kinetic energy is proportional to $\dot{x}^2$ and the potential energy to x^2. Each of these contributes $\frac{1}{2}kT$ to the average thermal energy.

Summarizing we may write: thermal energy per degree of translational freedom $= \frac{1}{2}kT$, per degree of rotational freedom $= \frac{1}{2}kT$ and per vibrational mode (equivalent to two degrees of freedom) $= kT$.

85 Further theory of perfect gases

Internal energy of a gas

If we consider the energy which arises solely from thermal motion (not excited electronic states), the internal energy u per molecule of a gas may be deduced for monatomic, diatomic and polyatomic molecules.

Monatomic gas: only 3 translations $u = \frac{3}{2} kT$ per molecule.

Diatomic gas: 3 translations $\left.\right\}$ $u = \frac{5}{2} kT$ per molecule plus any
          2 rotations   vibrational energy.

The third rotation about the axis of the molecule does not occur for reasons to be described later.

Polyatomic gas: 3 translations $\left.\right\}$ $u = \frac{6}{2} kT$ plus any vibrational energy.
            3 rotations

These results are summarized in Table 14.

Table 14 *Internal energy and specific heats of gases*

Gas	Monatomic	Diatomic	Polyatomic
f_{tr}	3	3	3
f_{rot}	0	2	3
f_{vib}	0	assumed 0	assumed 0
Total	3	5	6
u mol^{-1}	$\frac{3}{2}kT$	$\frac{5}{2}kT$	$\frac{6}{2}kT$
U g-mole^{-1}	$\frac{3}{2}RT$	$\frac{5}{2}RT$	$\frac{6}{2}RT$
$C_V = \left(\dfrac{\partial U}{\partial T}\right)_V$	$\frac{3}{2}R$	$\frac{5}{2}R$	$3R$
$C_P = C_V + R$	$\frac{5}{2}R$	$\frac{7}{2}R$	$4R$
$\gamma = \dfrac{C_P}{C_V}$	$\frac{5}{3}$	$\frac{7}{5}$	$\frac{4}{3}$
C_V cal. g-mole^{-1}:			
Theoretical	3	5	6
Observed	Ar: 3	H_2 : 4·9	H_2S : 6
	He: 3	O_2 : 5·0	CS_2 : 10
		N_2 : 4·6	
		Cl_2 : 6	

For monatomic gases the agreement is excellent. For diatomic gases the agreement is good, but the value given for chlorine seems to indicate that it possesses additional vibrational energy at room temperature. With poly-atomic molecules the behaviour is very complicated but it would seem that, in general, $C_V > 6$ calorie per gram-molecule.

In general if one knows the number of degrees of freedom f which are 'active' one can write

$$U = f \times \frac{RT}{2},$$

so that $C_V = \frac{1}{2}Rf$

and $\qquad \gamma = \frac{C_P}{C_V} = 1 + \frac{2}{f}.$ (5.56)

5·5·3 *Gas kinetic theory of adiabatic expansion*

Before we go on to discuss in greater detail the vibrational and rotational energy of a molecule, we may at once use the ideas expressed in the previous section to deduce the adiabatic gas equation for monatomic and polyatomic gases.

In an adiabatic expansion all the energy is taken from the internal energy of the gas. We first consider a monatomic gas where the internal energy arises solely from the kinetic energy of translation.

From the kinetic theory we have:

$$PV = \tfrac{2}{3} \text{ (total kinetic energy } Z)$$

$$Z = \tfrac{3}{2}PV = \text{internal energy } U.$$

The work done by the gas in an adiabatic expansion is

$$PdV = -dU = -dZ = -d(\tfrac{3}{2}PV).$$

$$\therefore PdV = -\tfrac{3}{2}PdV - \tfrac{3}{2}VdP$$

or $\qquad \tfrac{5}{2}PdV + \tfrac{3}{2}VdP = 0$

or $\qquad 5\dfrac{dV}{V} + 3\dfrac{dP}{P} = 0.$

Hence $\quad PV^{\frac{5}{3}} = constant.$ (5.57)

For a diatomic gas possessing two rotational degrees of freedom, the internal energy is

$$U = \text{kinetic energy} + \text{rotational energy}$$

$$= \tfrac{3}{2}PV + RT = \tfrac{3}{2}PV + PV$$

$$= \tfrac{5}{2}PV.$$

87   Further theory of perfect gases

For an adiabatic expansion $PdV = -d(\frac{5}{2}PV)$. This leads, as above, to

$$PV^{7/5} = \text{constant.} \tag{5.58}$$

5·5·4 Rotational and vibrational degrees of freedom

At room temperature the specific heat C_V of hydrogen is about 5 cal. mole^{-1}. As the temperature is lowered C_V decreases until at $-220°$C., C_V has a value of only about 3 cal. mole^{-1}. It is apparent that the rotational degrees of freedom have gradually disappeared and that at about 50°K. the molecule has only translational energy. The reason for this is that according to the quantum theory rotors can possess only discrete energy levels. For a symmetrical diatomic molecule rotating about an axis, for which the moment of inertia is I, the possible energy levels are

$$E_r = n(n+1)\frac{h^2}{8\pi^2 I}, \tag{5.59}$$

where h is Planck's constant ($6\cdot6 \times 10^{-27}$ erg sec.$^{-1}$) and n is an integer. The smallest value of rotational energy permitted occurs when $n = 1$ and has the value

$$(E_r)_{\text{min}} = \frac{h^2}{4\pi^2 I}. \tag{5.60}$$

For a hydrogen molecule, I about an axis normal to the bond is approximately $4\cdot7 \times 10^{-41}$ g. cm.^{2}

$$\therefore \ (E_r)_{\text{min}} = 2 \times 10^{-14} \text{ erg.}$$

But the translational energy $E_t = \frac{3}{2}kT = 2 \times 10^{-16}T$ erg, so that at 50°K.

$$E_t = 10^{-14} \text{ erg.} \tag{5.61}$$

Therefore the translational energy is appreciably less than the lowest rotational energy level. By Boltzmann's principle the fraction of molecules capable of acquiring rotational energy 2×10^{-14} erg at 50°K. will be

$$\exp\left(-\frac{\Delta\varepsilon}{kT}\right) \simeq e^{-3} \simeq \frac{1}{20}. \tag{5.62}$$

Consequently below 50°K. very few of the molecules will take up rotational motion. At room temperature (300°K.), kT will be large compared with the first rotational energy level, so a large fraction of the molecules will be able to take up the first and higher rotational energy levels.

Equation (5.60) also shows why rotating about the axis itself does not occur. The moment of inertia I is so small that the lowest rotational energy state, $\dfrac{h^2}{4\pi^2 I}$, is enormous and can never be reached before thermal dissociation occurs.

Similar considerations apply to the vibrational degrees of freedom. If the vibrational frequency is v, the first energy state (ignoring zero-point energy) demands energy of amount equal to hv. For the hydrogen molecule $v \approx 2{\cdot}6 \times 10^{14}$ sec.$^{-1}$, so that hv is about 2×10^{-12} erg. For an appreciable fraction of molecules to be excited to this level kT must have about the same value, so that T must exceed $10{,}000°$C. The molecule dissociates long before this. By contrast the chlorine molecule which has a much lower natural frequency v, is able to take up appreciable vibrational energy at temperatures above $600°$C. so that its specific heat approaches $C_V = 7$ cal. g-mol.$^{-1}$ at elevated temperatures (see figure 28).

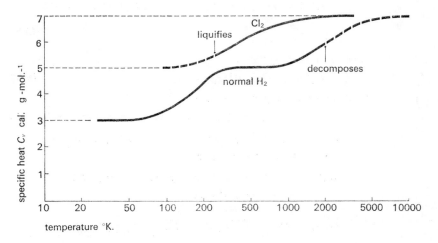

Figure 28.   Variation of specific heat with temperature for H_2 and for Cl_2.

5·5·5   Calculation of vibrational energy as a function of temperature

We reproduce here the standard treatment for a simple oscillator of natural frequency v. According to quantum theory the possible energy levels are

$$E_v = (n+\tfrac{1}{2})hv, \tag{5.63}$$

where n is an integer.

The zero-point energy $\tfrac{1}{2}hv$ is the vibrational energy the molecule possesses even at absolute zero, but since this is a constant we may ignore it at this stage of the computation and add it at the end. We put

$$E_v = nhv. \tag{5.64}$$

89   Further theory of perfect gases

Applying the Boltzmann function, the number of states with quantum number n is

$$N_n = A \exp -\frac{nhv}{kT}$$

$$= A \exp(-n\beta) \quad \text{where} \quad \beta = \frac{hv}{kT}. \tag{5.65}$$

The total number N of atoms is the sum of all N_ns from $n = 0$ to $n = \infty$,

$$N = A[1 + \exp(-\beta) + \exp(-2\beta) + \exp(-3\beta) + \ldots]$$
$$= A[1 + \exp(-\beta) + (\exp(-\beta))^2 + (\exp(-\beta))^3 + \ldots]$$
$$= \frac{A}{1 - \exp(-\beta)}. \tag{5.66}$$

This gives A so that N_n becomes

$$N_n = N[1 - \exp(-\beta)][\exp(-n\beta)]. \tag{5.67}$$

The total energy is the sum of $N_n E_v$ where $E_v = nhv$.

$$U = N[1 - \exp(-\beta)][\exp(-n\beta)] nhv \text{ summed for all values of } n$$
$$= N[1 - \exp(-\beta)] \{hv[\exp(-\beta)] + 2hv[\exp(-\beta)]^2 +$$
$$+ 3hv[\exp(-\beta)]^3 + \ldots\}$$
$$= N[1 - \exp(-\beta)] hv[\exp(-\beta)] \{1 + 2[\exp(-\beta)] +$$
$$+ 3[\exp(-\beta)]^2 + \ldots\}$$
$$= N[1 - \exp(-\beta)] hv[\exp(-\beta)] \frac{1}{[1 - \exp(-\beta)]^2}$$
$$= Nhv \frac{[\exp(-\beta)]}{[1 - \exp(-\beta)]} = \frac{Nhv}{\exp(\beta) - 1}. \tag{5.68}$$

If we include the zero-point energy $\frac{1}{2}hv$ we have

$$U = Nhv\{\tfrac{1}{2} + [\exp(\beta) - 1]^{-1}\}. \tag{5.69}$$

At high temperatures $\exp \beta = \exp \dfrac{hv}{kT} \simeq 1 + \dfrac{hv}{kT}$ so that

$$U \simeq Nhv \left[\frac{1}{2} + \frac{kT}{hv}\right]$$

$$U = \tfrac{1}{2}Nhv + NkT. \tag{5.70}$$

Thus apart from zero-point energy each vibration carries average energy kT, so that the specific heat contribution is $Nk = R$ for each vibration. At lower temperatures $\left[\exp\left(\dfrac{hv}{kT}\right) - 1\right]^{-1}$ and therefore the specific heat falls off and tends to zero when $T = 0$ (see figure 29). This decrease in vibrational energy with temperature reproduces essentially the right-hand features of the specific heat curves of hydrogen and chlorine in figure 28. We may also note

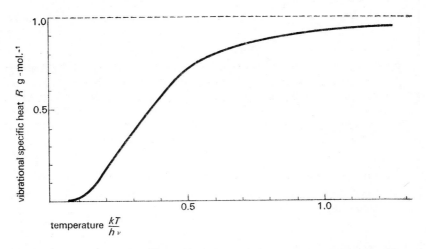

Figure 29. Variation of vibrational specific heat with temperature for a simple oscillation of natural frequency v. Specific heat is in units of R per g-mole; temperature in units of $\dfrac{k}{hv}$.

in passing that this concept was applied by A. Einstein to explain the specific heat of solids. He suggested that each atom has 3 independent vibrational modes of constant frequency v. The specific heat (in this case it assumes no volume change and is therefore C_V) is $3R$ at high temperatures and falls to zero at low temperatures. We shall discuss this in greater detail in a later chapter.

5·5·6   Rotational energy

For a symmetrical diatomic molecule rotating about an axis for which the moment of inertia is I, the possible energy levels E_r are as given in equation (5.53). The application of the Boltzmann distribution is complicated because each state of quantum number n consists of g states of practically identical energy. The degeneracy factor g has the value

$$g = 2n + 1. \tag{5.71}$$

One can build up the Boltzmann relation as before where the basic exponential term is $\exp\left(-\dfrac{E_r}{kT}\right)$ or $\exp\left(-n(n+1)\dfrac{h^2}{8\pi^2 I}\dfrac{1}{kT}\right)$. The final integral has to be summed numerically. For high temperatures the energy for N rotors becomes NkT. This is the classical result for a rotor with two degrees of freedom. In general the rotational energy falls off from NkT when

$$kT < \frac{h^2}{8\pi^2 I}.\tag{5.72}$$

5·5·7 Translational energy

For a gas confined within a container, quantization of translational energy does occur, but the number of energy-states is enormous and the spacing between energy levels is minute. We may therefore consider all energy states to be available as a continuous distribution, so that the classical Boltzmann distribution applies. Consequently each degree of translational freedom involves energy $\frac{1}{2}kT$ per molecule under all practical conditions.

5·6 Macroscopic examples of equipartition of energy

We conclude this chapter with four examples of thermal motion occurring in systems other than gases.

5·6·1 Galvanometer mirror

A delicately suspended galvanometer mirror can be seen to undergo small random oscillations which are due to its thermal energy. The mirror has a single degree of oscillation about its axis so that we attribute to its mean square velocity $(\overline{\omega^2})$, energy $\frac{1}{2}kT$, and to its mean square angular deflection $(\overline{\theta^2})$, energy $\frac{1}{2}kT$. We may write

$$\tfrac{1}{2}kT = \tfrac{1}{2}I(\overline{\omega^2}) = \tfrac{1}{2}\tau\,(\overline{\theta^2}),\tag{5.73}$$

where I is the amount of inertia, τ is the torsion constant. If we determine the torsional stiffness of the suspension we may then calculate $(\overline{\theta^2})$. This tells us the amount by which any galvanometer reading will fluctuate as a result of thermal motion. These fluctuations do not depend on the presence of air surrounding the mirrors; it is true that gas molecules will buffet the mirror and contribute to the oscillation, but they will also contribute to the damping, so the net effect is unchanged. The mirror behaves like a large molecule with a single mode of torsional oscillation. In a perfect vacuum the oscillations can be considered as arising from the random absorption and radiation of electromagnetic energy associated with the temperature T. The resulting behaviour is the same.

5·6·2 Sedimentation

As we saw earlier particles in thermal equilibrium behave like gas molecules of large molecular weight. If they are in an evacuated container the sedimentation due to gravity is so rapid that they virtually all lie at the bottom of the container. If, however, they are suspended in a liquid of very nearly the same density the effective gravitational field is greatly reduced and there may be an appreciable variation in particle concentration with distance from the bottom of the container as given by equation (5.35).

5·6·3   *Electrical noise*

The conduction electrons in a metal may be regarded as a gas with random velocities. These give rise to electrical 'noise'. The root mean square potential fluctuation across a resistor of resistance Ω is given by

$$\text{r.m.s. potential} = \sqrt{[(4\Omega kT(f_2 - f_1)]},\tag{5.74}$$

where $f_2 - f_1$ is the bandwidth frequency over which the measurements are made. This result at once shows that there is a limit to useful amplification. If the original signal is too small (e.g. comparable with the random electrical noise) no amount of amplification will improve reception.

Similarly in the thermionic emission of current I_0 it may be shown that there is a fluctuation i in the emitted current given by

$$|\overline{i^2}|^{\frac{1}{2}} = \sqrt{\left(\frac{I_0 e}{\tau}\right)},\tag{5.75}$$

where e is the electronic charge and τ the time constant of the measuring system. This effect has indeed been used as an independent means of determining e; the result agrees with other methods to within 1 per cent.

5·6·4   *Brownian motion*

The English botanist R. Brown observed in 1827 that finely divided particles in suspension undergo ceaseless movement. This motion was analysed by Einstein in 1906; the following derivation is based on a treatment developed by P. Langevin in 1908. We first assume that the particles possess thermal kinetic energy in one dimension of amount $\frac{1}{2}kT$.

If at any instance the molecules of the liquid collide with an individual particle to produce a net force X in the x direction, we may write the equation of motion of the particle in the form

$$m\ddot{x} + 6\pi a\eta\dot{x} = X,\tag{5.76}$$

where m is the mass, a the radius of the particle (assumed spherical) and η the viscosity of the fluid. In this equation the first term represents the inertial force on the particle and the second the frictional force due to the viscous resistance of the fluid; this is Stokes' well-known relation for viscous forces on a sphere. The third term represents the force due to molecular collisions. Thus from the point of view of collisions we treat the fluid as a collection of individual molecules, but from the point of view of the resistive forces we

93   Further theory of perfect gases

treat the fluid as a continuum. Finally this equation refers only to motions of the particle produced by collisions; the intrinsic motion of the particle by virtue of its own kinetic energy ($\frac{1}{2}kT$) emerges from the analysis.

We multiply both sides of equation (5.76) by x and then average over a large number of collisions. Then for any fixed specified value of x, for every value of X there is an equal chance of $-X$ occurring so that $\Sigma xX = 0$.

$$\Sigma m\ddot{x}x + 6\pi a\eta \Sigma \dot{x}x = \Sigma xX = 0 \qquad (5.77)$$

or
$$m\,\overline{\ddot{x}x} + 6\pi a\eta\,\overline{\dot{x}x} = 0.$$

We note that $\dfrac{d}{dt}(\dot{x}x) = \ddot{x}x + \dot{x}^2 = \ddot{x}x + u^2$ where u is the velocity in the x direction. Then

$$m\frac{d}{dt}\,\overline{\dot{x}x} - m\overline{u^2} + 6\pi a\eta\,\overline{\dot{x}x} = 0. \qquad (5.78)$$

It turns out that the first term implies a transient at the beginning of the collision which has a negligible effect on the subsequent displacement.*
Hence
$$6\pi a\eta\,\overline{\dot{x}x} = m\overline{u^2} = kT, \qquad (5.79)$$

or
$$3\pi a\eta\,\frac{d}{dt}(\overline{x^2}) = kT. \qquad (5.80)$$

In a time interval τ the mean square displacement $\overline{\Delta x^2}$ is then

$$\overline{\Delta x^2} = \frac{kT\tau}{3\pi a\eta} = \frac{RT\tau}{3\pi a\eta N_0}, \qquad (5.81)$$

where N_0 is Avogadro's number. A given particle is observed say every 30 seconds and the displacement Δx in the x direction is observed for each interval of time. These values are squared and the mean value is calculated; this gives $\overline{\Delta x^2}$. This is an average path for the time interval $\tau = 30$ seconds. The value of $\overline{\Delta x^2}$ so found may be compared with the right-hand term of

---

* More precisely we may write $\dot{x}x$ as $\dfrac{1}{2}\dfrac{d}{dt}(x^2)$ which we may call z. Equation (5.78) then becomes
$$m\dot{z} + 6\pi a\eta z = m u^2 = kT.$$
The solution is
$$z = \frac{kT}{6\pi a\eta} + A\exp\left(\frac{-6\pi a\eta t}{m}\right),$$
where A is a constant of integration. The quantity $\dfrac{6\pi an}{m}$ is extremely large so that even for minute values of t the exponential term becomes negligible. During the total time of the collision, z is thus given essentially by the first term on the right-hand side of the equation. This is indeed the same equation as (5.80).

equation **(5.81)**. Experiments show that the results agree very well with calculation.

There are two points of interest. First the displacement does not depend on the mass of the particle (only on its radius). Secondly although the individual particle has intrinsic motion, which has nothing to do with the surrounding liquid, the distance it moves about an equilibrium position *is* determined by the viscous damping of the liquid. In a descriptive way we may say that a viscous force $6\pi a\eta\dot{x}$ acting over a distance x consumes the thermal energy $\frac{1}{2}mu^2$. The particle then reacquires its thermal energy and the process is repeated. Apart from a numerical factor this is essentially the result implied by equation **(5.79)**.

If then equation **(5.79)** is a description of the way in which viscous damping consumes thermal energy, is it not possible to apply it to the behaviour of a gas molecule within its own gas? We have already derived a relation for the viscosity of a gas, namely $\eta = \frac{1}{3}mn\bar{c}\lambda$. However this can only be used when the distances between the moving bodies is larger than the mean free path λ. This implies that Stokes' law is valid for gases only if the radius of the sphere is larger than λ. Thus we cannot really use Stokes' law if the sphere is the gas molecule itself. However as an extreme case let us see what happens if we take a heavily condensed gas where $\lambda = a$. We have

$$\eta = \tfrac{1}{3}mn\bar{c}a; \quad \dot{x} \simeq u; \quad x \simeq a.$$

Then the L.H.S. of equation **(5.79)** becomes

$$6\pi a\eta\dot{x}x = 6\pi a\tfrac{1}{3}mn\bar{c}.u.a = 2\pi a^3 nm\bar{c}u.$$

But $\qquad\bar{c} \simeq \sqrt{3}(u).$

Hence $\quad 6\pi a n\dot{x}x \simeq 10a^3 nmu^2.$ **(5.82)**

But each sphere of radius a occupies a cube of volume $8a^3$, so that $8a^3 n = 1$. The value of equation **(5.82)** is thus approximately mu^2, which agrees with the r.h.s. of equation **(5.79)**. This is, of course, pushing the model to an extreme, but it shows again that the particle considered in the analysis of Brownian motion is essentially no different from a molecule, except that it is very much larger.

This has been a long chapter, and we now conclude it with a relevant quotation from Lucretius:

This process, as I might point out, is illustrated by an image of it that is continually taking place before our very eyes. Observe what happens when sunbeams are admitted into a building and shed light on its shadowy places. You will see a multitude of tiny particles mingling in a multitude of ways in the empty space within the light of the beam, as though contending in everlasting conflict, rushing into battle rank upon rank with never a moment's pause in a rapid sequence of unions and dis-unions. From this you may picture what it is for the atoms to be perpetually tossed about in the illimitable void. To some extent a small thing may afford an illustration and an imperfect image of great things. Besides, there is a further reason

95 Further theory of perfect gases

why you should give your mind to these particles that are seen dancing in a sunbeam: their dancing is an actual imitation of underlying movements of matter that are hidden from our sight. There you will see many particles under the impact of invisible blows changing their course and driven back upon their tracks, this way and that, in all directions. You must understand that they all derive this restlessness from the atom. It originates with the atoms, which move themselves. Then those small compound bodies that are least removed from the impetus of the atoms are set in motion by the impact of their invisible blows and in turn cannon against slightly larger bodies. So the movement mounts up from the atoms and gradually emerges to the level of our senses, so that those bodies are in motion that we see in sunbeams, moved by blows that remain invisible.

(It is interesting to note that, almost two thousand years later Perrin used the identical argument in his studies of Brownian motion of gamboge particles in a liquid: he suggested that this could be projected optically before an audience to demonstrate the molecular movement within the liquid.)

Prof. S. Sambursky in his perceptive study *The Physical World of the Greeks* adds the comment that this remarkable description, 'perfectly describes and explains the Brownian movement by a wrong example. The movements of dust particles as seen by the naked eye in sunlight are caused by air-currents: the real phenomenon postulated by Lucretius on the basis of abstract reasoning can only be seen in a microscope.' Further, as we have pointed out previously, the random motion of fine particles is an intrinsic property of any assembly of free entities in thermal equilibrium, whether they be gas molecules or solid particles. Nothing, however, can detract from the beauty and elegance of this inferential type of argument.

Integrals

Values of $\int_0^\infty x^n e^{-ax^2} dx$:

$$\int_0^\infty e^{-ax^2} dx = \frac{1}{2}\sqrt{\frac{\pi}{a}}$$

$$\int_0^\infty x e^{-ax^2} dx = \frac{1}{2a}$$

$$\int_0^\infty x^2 e^{-ax^2} dx = \frac{1}{4}\sqrt{\frac{\pi}{a^3}}$$

$$\int_0^\infty x^3 e^{-ax^2} dx = \frac{1}{2a^2}$$

For all these integrals $\int_{-\infty}^\infty = 2\int_0^\infty$

Chapter 6
Imperfect gases

Many gases even at room temperature deviate markedly from the 'ideal' gas relation $PV = RT$. Similar deviations are observed with noble gases at higher pressures and lower temperatures. In this chapter we shall discuss the behaviour of such 'imperfect' gases and show that their properties can be explained in terms of the 'ideal' gas theory, modified to allow for the finite volume of the gas molecules and for the existence of intermolecular forces.

6·1 Deviations from perfect gas behaviour

6·1·1 The virial equation

In general, instead of the relation PV = constant at a fixed temperature, imperfect gases obey an empirical power equation of the form

$$PV = A + BP + CP^2 + DP^3. \tag{6.1}$$

This is known as a virial equation and A, B, C, and D as virial coefficients. The most important coefficient, apart from A, is B, since C and D are usually very small. At low temperatures B is negative, at higher temperatures positive and at some intermediate temperature T_B it has the value zero. At this temperature, if the pressures are not too high we see at once that equation (6.1) reduces to

$$PV = A = \text{constant}, \tag{6.2}$$

that is Boyle's law becomes approximately true again. For this reason T_B is called the Boyle temperature.

Plots of PV against P are shown schematically in figure 30. If we measure the slope of the curves near the PV axis, i.e. near $P = 0$ we have

$$\frac{\partial(PV)}{\partial P} = B + 2CP + \ldots$$

$$= B \text{ for } P \rightarrow 0. \tag{6.3}$$

The slope near $P = 0$ will, therefore, be zero at the Boyle temperature. It is also easy to show that the curve connecting the minima is a parabola.

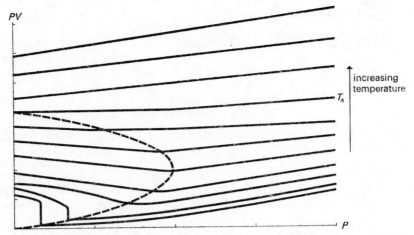

Figure 30. Variation of PV with P for a real gas. At one special temperature where the initial tangent is horizontal PV is constant at the lower pressure range; this is the Boyle temperature T .

6·1·2 *Andrews' experiments on carbon dioxide*

T. Andrews (1813–1885) carried out a series of important and classical experiments on the compressibility of gases and the conditions under which they can be liquefied. His first paper appeared in 1861 and one of his major surveys was given in his Bakerian lecture of 1869. We summarize his main conclusions as exemplified by the behaviour of CO_2 (see figure 31).

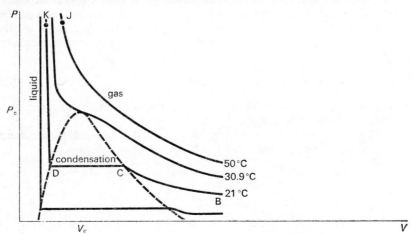

Figure 31. Variation of pressure P with volume V for CO_2 (based on Andrews' results).

98   Gases, liquids and solids

(i) Above a temperature of about 48°C., the compressibility of CO_2 resembles that of a perfect gas.

(ii) At, say, 21°C. compression produces liquefaction.

From B to C the material behaves as a gas, from C to D it condenses at a constant pressure, and at D it is completely liquid. From D to E the slope is very steep because a liquid is relatively incompressible.

(iii) At 31·1°C. there is no liquefaction although the isotherm is very distorted from the ideal gas isotherm.

(iv) At 30·9°C. liquefaction just occurs under compression.

Above this temperature, liquefaction cannot be produced. This is known as the critical temperature T_c, the temperature above which liquefaction cannot be achieved however high the pressure. The pressure at which liquefaction just occurs is called the critical pressure P_c, and the corresponding volume (for one gram-molecule of the substance) is called the critical volume V_c.

6·1·3 Continuity of liquid and gaseous state

Let us consider two isothermals, one above T_c and one below. Consider the points J and K on these isothermals at some pressure level above P_c. At J the substance is purely a gas, at K purely a liquid; this means that if we keep the pressure constant at J but steadily reduce the temperature we may pass to the liquid phase at K. This, in turn, implies that above P_c it is possible, simply by cooling, to pass from the gaseous to the liquid state without any mixture of phases.

6·1·4 Significance of T_c

The critical temperature may be easily understood in terms of intermolecular forces. Suppose figure 32 represents the potential energy between one molecule

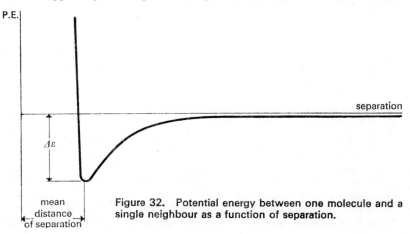

Figure 32. Potential energy between one molecule and a single neighbour as a function of separation.

99 Imperfect gases

and a single neighbour, as a function of separation. It is clear that if the thermal energy of the molecule is greater than $\Delta\varepsilon$, one molecule can always escape from its neighbours. However closely they are pressed together (within limits) the thermal energy will always be sufficient to overcome the attraction between the molecules.

Now the thermal energy of each molecule is of order kT where k is Boltzmann's constant and equals $1\cdot4 \times 10^{-16}$ erg. deg.$^{-1}$ and T is the absolute temperature. Thus a molecule can always escape from its neighbour if $kT > \Delta\varepsilon$

or $\qquad T > \dfrac{\Delta\varepsilon}{k}$. $\hfill$ (6.4)

The substance is therefore unliquefiable for temperatures above

$$T_c = \frac{\Delta\varepsilon}{k}.$$

Some typical results are given in Table 15. It is seen that the agreement is good, indeed surprisingly good, in view of the extremely simple assumptions made. This model deals only with the problem of a molecule escaping from a

Table 15 *Critical temperatures*

Gas	$\Delta\varepsilon$ ergs (theoretical)	$\dfrac{\Delta\varepsilon}{k}$ °K.	T_c (observed) °K.
He	$0\cdot8 \times 10^{-15}$	6	5·2
H_2	4×10^{-15}	30	33·1
N_2	13×10^{-15}	100	121

single neighbour. In section 6·2·6·2 we shall consider this behaviour in somewhat greater detail and, using a different model, establish essentially the same results.

6·1·5 *The miscibility of liquids and gases*

The continuity of the liquid and gaseous states has an interesting bearing on the solubility or miscibility of liquids. Our starting point is that, unlike solids and liquids, all gases are completely soluble in one another (see p. 46). Let us now consider the behaviour of, say, water and oil: these are mutually insoluble in the liquid state and if mixed together will separate into two distinct phases. However, water vapour and hydrocarbon vapour form a homogeneous mixture in the gaseous phase, in which there is a perfect random mixing of molecules of the two materials. If now the vapour is very highly compressed it can be brought to a density of the ordinary liquid; but provided the temperature is high enough – above the 'effective critical temperature' of the system – it will behave as a gas and still remain as a single-phase material. We shall thus have achieved complete miscibility of oil in water. Of course, this can be seen from a completely different angle. If water and oil are enclosed in a container at constant volume and heated, a temperature will be reached at which the thermal energy is sufficient to overcome the intermolecular attraction of like and unlike molecules – the material will then cease to consist of two separate phases and complete miscibility will have been achieved Alternatively we may say that by increasing the temperature the entropy of the system is increased and the randomization which this implies may be sufficient to mix the materials on a molecular scale, i.e. to achieve miscibility. Is the single-phase material now a high-density gas or a high-temperature liquid? To some extent this is a matter of semantics.

6·2 Kinetic theory of an imperfect gas: van der Waals equation

6·2·1 *Derivation of van der Waals equation*

We shall now derive an equation for an imperfect gas, using the assumption that the main deviations from ideal behaviour arise from the finite size of the molecules and the forces between them.

We first assume that if real and ideal gases are in thermal equilibrium, that is at the same temperature, the mean translational energy of the individual gas molecules ($\frac{1}{2}m\overline{c^2}$) still implies a value of $\overline{c^2}$ given by $\frac{3RT}{M}$. This assumption, in effect, is a broader generalization of the discussion given on page 45. (Note that the Boltzmann distribution law, $n_\varepsilon = N \exp\left(-\frac{\varepsilon}{kT}\right)$, applies only if ε includes both kinetic and intermolecular potential energy.)

Size of molecule. We first deal with the finite size of the molecules. We may say that the collisions with the wall of the container which determine the observed gas pressure are exactly the same as for zero-size molecules, i.e. an ideal gas, but that the available volume is reduced by some amount b depending on the number and size of the molecules. We may, therefore, write for one gram-molecule

$$P(V-b) = RT. \tag{6.5}$$

There are several ways of estimating b.

(i) Molecular collisions between the walls. If σ is the effective diameter of a molecule $\left(\text{volume } v_m = \frac{\pi\sigma^3}{6}\right)$ the mean free path between collisions is reduced from λ to $\lambda - \sigma$. Thus as the molecules travel between the walls of the vessel they will travel $(\lambda - \sigma)$ between collisions where, on the ideal gas model, they would have to travel a distance λ. Thus the number of collisions on the wall per second is increased in the ratio $\dfrac{\lambda}{\lambda - \sigma}$. This means that the pressure is increased by the factor $\dfrac{1}{1 - \left(\dfrac{\sigma}{\lambda}\right)}$. Instead of the ideal gas equation

$$PV = \tfrac{1}{3} Nm\overline{c^2} = RT,$$

we therefore have

$$PV = \frac{RT}{\left(1 - \dfrac{\sigma}{\lambda}\right)} \qquad (6.6)$$

or

$$P\left(V - V\frac{\sigma}{\lambda}\right) = RT. \qquad (6.7)$$

The correction term b in equation (6.5) therefore has the value

$$b = V\frac{\sigma}{\lambda} = V\sigma(\pi\sigma^2 n)$$

$$= Vn\pi\sigma^3$$

$$= N_0 6 v_m, \qquad (6.8)$$

where we have used the simple value of $\lambda = \dfrac{1}{\pi\sigma^2 n}$ derived in Chapter 4. On this model, therefore, b is about 6 times the total volume occupied by the molecules.

(ii) Collisions with the wall (Jeans). Jeans gives a rather tricky way of calculating b. He suggests that all molecules are surrounded by a sheath of radius σ which excludes the other molecules. The excluded volume is therefore 8 times the volume of the molecules themselves. For a gram-molecule containing N_0 molecules this is therefore 8 $N_0 v_m$. There is thus a zero chance of finding a molecule within the excluded volume and a chance

$$\frac{dV}{V - 8 N_0 v_m}, \qquad (6.9a)$$

of finding it in an element of volume dV outside the excluded volume. He then considers the space available for molecules about to collide with the wall and argues that, since only the farther halves of the spherical sheaths are excluded when collisions are imminent, the chance of dV being free in space is

$$\frac{V-4N_0 v_m}{V}.$$ (6.9b)

The chance of a molecular collision with the wall is found by multiplying the two probabilities together. The result is

$$\frac{dV}{V} \times \frac{V-4N_0 v_m}{V-8N_0 v_m} \simeq \frac{dV}{V} \frac{V}{V-4N_0 v_m}.$$

For a gas of negligible size the probability is $\dfrac{dV}{V}$ so that for the real gas the collision probability, and hence the pressure, are increased by the ratio $\dfrac{V}{V-4N_0 v_m}$. This is equivalent to replacing V by $V-4N_0 v_m$ in the ideal gas equation so that

$$b = 4N_0 v_m.$$

Readers who find it difficult to appreciate the subleties of this treatment may prefer the following.

(iii) Excluded volume. If a molecule approaches another within a distance σ between centres contact occurs. Thus each molecule appears to carry a sheath of radius σ, i.e. of volume $8v_m$ which excludes other molecules. If, therefore, we start with a container of volume V, the first molecule has a free volume V; the second a free volume of $V-8v_m$; the third a free volume of $V-2 \times 8v_m$; the fourth a free volume of $V-3 \times 8v_m$. We then have to find an effective average value for all the N molecules in the gas. The standard procedure is to take the geometric mean, i.e. the Nth root of the product

$$V(V-8v_m)(V-2 \times 8v_m)(V-3 \times 8v_m) \dots$$

One thus obtains as the average available volume a value very close to

$$V-N_0 4v_m,$$

so that $b = 4N_0 v_m.$ (6.10)

(iv) Excluded volume simplified. A much simpler approach (no less valid) is to say that each molecule carries an excluding volume of $8v_m$. But if we are concerned only with double collisions (ignoring triple and higher collisions) this means that on the average $8v_m$ is the excluded volume for a pair of molecules. The average is $4v_m$ so that once again we have

$$b = 4N_0 v_m.$$ (6.11)

We have given these four different approaches to show that there is a simple numerical factor relating the excluded volume to the volume of the molecule but that its value depends on the model used to derive it. Experiments agree well with a value of about 4 but readers should realize that this is a 'good' value, not a 'correct' value.

6·2·2   *Intermolecular forces*

We now turn to the effect of intermolecular forces. Within the bulk of the gas the molecular forces, on an average, act symmetrically on one another so that the net effect is zero. Consequently within the bulk of the gas the molecules behave as though they were in a gas without attraction, so that the effective pressure is the same as for an ideal gas P_i. Near the walls of the vessel, however, the molecules have to escape from their neighbours before they collide with the wall. In overcoming this molecular attraction some kinetic energy is lost, the molecular velocity is reduced and the momentum imparted to the wall on impact and rebound is less than it would be for an ideal gas. Suppose the pressure defect is ΔP. Then the observed pressure P at the walls is less than P_i by an amount ΔP.

If we write the ideal gas equation, modified to allow for this, we have

$$P_{\text{ideal}}\, V_{\text{ideal}} = RT,$$

or $$(P + \Delta P)(V - b) = RT. \tag{6.12}$$

The pressure defect is proportional to the number of molecules striking the surface per unit area per second, i.e. to the density ρ. It is also proportional to the number of molecules attracting the molecules and reducing their impact. This is also proportional to ρ. As a result ΔP is proportional to ρ^2. It could be written as aP^2 since P is nearly proportional to ρ, but it turns out that it is better to write this as $\dfrac{a}{V^2}$; this correction term appearing to operate over a larger range than a term aP^2. We may thus regard the correction term $\dfrac{a}{V^2}$ as the internal pressure in the gas, i.e. the pressure that has to be exerted to pull the molecules away from one another in overcoming intermolecular forces. Our final relation is

$$\left(P + \frac{a}{V^2}\right)(V - b) = RT. \tag{6.13}$$

This celebrated relation is known as the van der Waals equation (published in 1873) and is a very convenient way of describing the behaviour of a 'real' gas. The model on which it is derived is admittedly a very crude one but it is generally felt that the equation holds better than it ought. Its main merit is that it is simple and that it allows for molecular volume and molecular attraction with the use of only two parameters a and b.

If a and b are determined for a gram-molecule of gas, the van der Waals equation of state for n gram-molecules becomes

$$\left(P+n^2\,\frac{a}{V^2}\right)(V-nb) = nRT.$$

6·2·3 *Other equations of state*

There are a number of other equations of state which at various times have been proposed. R. Clausius (1822–1888) attempted to allow for the effect of temperature on the cohesive forces and suggested in 1880,

$$\left(P+\frac{a}{T(V+c)^2}\right)(V-b) = RT, \tag{6.14}$$

but this involves three arbitrary constants. D. Berthelot in 1878 found that if he used a different value for a he could get about as good a fit with experiment by putting $c = 0$. He therefore used

$$\left(P+\frac{a}{TV^2}\right)(V-b) = RT. \tag{6.15}$$

The best known 'rival' to the van der Waals equation of state is that due to C. Dieterici (published in 1899). The Dieterici treatment uses an excluded volume b as does the van der Waals equation. Its approach to the pressure defect due to intermolecular attraction is, however, different: it is, in some ways, more sophisticated. The van der Waals treatment emphasizes the diminution in momentum transfer because of the reduction in velocity of impact, whereas Dieterici emphasizes the reduction in number of molecules hitting the surface in unit time. The number of molecules to reach the wall in unit time with a specified velocity is less than the ideal number, because they need to possess a higher energy in order to escape from their neighbours. The energy needed to escape is proportional to the density of molecules n, so we may write

$$\text{number} = \text{ideal number} \times \exp\left(-\frac{\beta n}{RT}\right)$$

where β is a suitable constant. Clearly n is proportional to $1/V$, so that βn may be written as α/V where α is another constant. The observed pressure P may then be written

$$P = P_{\text{ideal}}\,\exp\left(-\frac{\alpha}{VRT}\right).$$

Since $P_{\text{ideal}}\,V_{\text{true}} = RT$ we have finally

$$P(V-b) = RT\exp\left(-\frac{\alpha}{VRT}\right). \tag{6.16}$$

105 Imperfect gases

It will be noted that apart from Clausius' relation all these equations involve only two arbitrary constants. As we shall see later this makes it possible to deduce a law of corresponding states for gases obeying the van der Waals, Berthelot or Dieterici equation.

6·2·4   Attraction of the walls

We have spent some space in discussing intermolecular attraction as the cause of a pressure 'defect' in real gases. What of the attraction between the molecules and the walls of the container? Does the pressure exerted by a gas depend on the nature of the container? Will the gas pressure be higher for a polar container than for an inert one? The answer is simple. The wall can have no net effect. Attraction may increase the impulse during the collision but the impulse will be reduced by an exactly equal amount due to the extraction of momentum from the wall. The net effect will be zero.

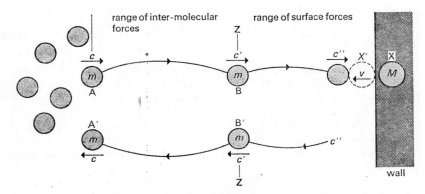

Figure 33.   Behaviour of a gas molecule as it escapes at A from the attraction of its neighbours and then approaches at B to within the range of action of the surface forces of the walls of the container.

This is shown schematically in figure 33. The velocity of a molecule in the bulk of the gas is, say, c. When it escapes from the attraction of its neighbours its velocity has fallen to c' (position B). At B it comes under the attractive forces of the wall. Consider the interaction with a wall molecule X of mass M. Suppose X is initially at rest. The molecule is accelerated towards X but also draws X out of the wall to X'. When collision occurs the velocity of the molecule (mass m) has been increased to c'' whilst the velocity of the wall molecule is, say, v. The gain in momentum of the molecule m $(c''-c')$ is exactly equal to the gain in momentum in the opposite direction Mv of the wall molecule. Thus the wall molecule is given an increased blow because of the attractive forces, but the increased momentum is exactly nullified by the momentum withdrawn from the wall. The net momentum transfer is the same as if no attraction had existed. On rebound the molecule leaves with increased velocity so imparting a larger momentum to the wall molecule, but as the gas molecule returns to position B′ the attractive forces decrease its velocity to its original value $c′$: at the same time they decrease the effective impulse on M, and it returns to the wall with the same momentum as though no attractive force had existed. The total momentum change is thus the same as if the wall exerted no force on the gas molecule. (This is simply a description of Newton's laws of momentum in terms of molecular interactions.) By the time the gas molecule reaches A′ its velocity is back to c and the bulk behaviour of the gas now resembles that of a perfect gas.

The problem discussed in this section is often avoided in elementary texts. Usually there is a bland statement that, by applying Newton's laws of momentum, the molecular momentum is reversed after collision with the wall. This is, of course, perfectly correct but it is equivalent to assuming that the wall exists at ZZ. It ignores the detailed interaction which takes place in the region between ZZ and the wall molecules themselves.

6·2·5   *Van der Waals equation and the Boyle temperature*

If we plot isothermals of PV against P we note that the Boyle temperature occurs when the slope of the curve is zero, at $P \to 0$. We therefore use the van der Waals equation, determine $\dfrac{\partial}{\partial P}(PV)$, and put it equal to zero at $P = 0$.

From the original equation (6.13) we have

$$P = \frac{RT}{V-b} - \frac{a}{V^2},$$

so that
$$PV = \frac{RTV}{V-b} - \frac{a}{V}. \tag{6.17}$$

Then

$$\frac{\partial}{\partial P}(PV) = \left\{ RT\frac{1}{V-b} - RT\frac{V}{(V-b)^2} + \frac{a}{V^2} \right\} \left(\frac{\partial V}{\partial P} \right)_T$$

$$= \left\{ -\frac{RTb}{(V-b)^2} + \frac{a}{V^2} \right\} \left(\frac{\partial V}{\partial P} \right)_T. \tag{6.18}$$

If we specify that, at $T = T_B$, this is zero when $P = 0$, this is equivalent to considering V as approaching infinity so that $(V-b)$ is indistinguishable from V. From equation (6.18) we therefore have

$$-RT_B\, b + a = 0,$$

or
$$T_B = \frac{a}{Rb}. \tag{6.19}$$

At all temperatures above T_B, PV will increase with increasing P, so that there will be no inflexion in the PV curves. At T_B itself we may substitute $T = T_B = \dfrac{a}{Rb}$ in the van der Waals equation. Multiplying out equation (6.13) we have

$$PV + \frac{a}{V} - bP - \frac{ab}{V^2} = RT.$$

On the left-hand side we may, if the pressures are small, replace $\dfrac{a}{V}$ by $\dfrac{aP}{RT}$, and neglect the fourth term. We are left with

$$PV + \frac{aP}{RT} - bP = RT.$$

From our value of the Boyle temperature $\left(T_B = \dfrac{a}{Rb}\right)$ we see that the second and third terms on the left-hand side are equal and opposite. Hence at T_B

$$PV = RT_B$$

for a real van der Waals gas.

6·2·6   *Van der Waals equation and the critical points*

The van der Waals equation is a cubic in V. If we plot P against V we obtain a family of curves shown schematically in figures 34 and 35. Let us consider a typical isotherm XDBACEY in figure 34 below the critical temperature. Join D and E by an isobar which cuts the curve at A.

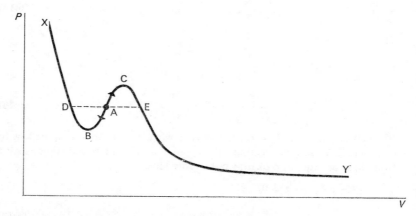

Figure 34.   Stylized behaviour of a van der Waals gas.

108   Gases, liquids and solids

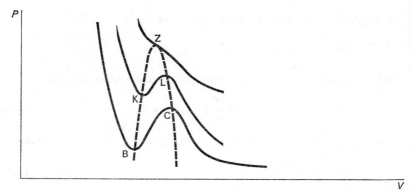

Figure 35. The van der Waals isotherms at various temperatures.

Consider now the behaviour at A. Suppose we started with a homogeneous phase. Along BC the slope $\frac{\partial P}{\partial V}$ is positive. This means that if a small increase in volume occurs this is accompanied by an *increase* in gas pressure. There is thus a spontaneous expansion towards C. Similarly a small decrease in volume would lead to a spontaneous contraction towards B. These processes constitute the initial separation into two phases, gaseous at C liquid at B. The part BD is unstable, the liquid can remain in this region (region of superheating), but generally the phase spreads to D. Similarly CE is unstable, the gas can remain in this region of super cooling but generally the phase spreads to E.* In passing we may note that some isotherms can give negative pressures i.e. B may be below the V axis. This only occurs in the liquid phase and as we shall see this provides a means of estimating the tensile strength of a liquid.

The two turning points B and C coalesce into a single region at Z on the critical isotherm (figure 35). We thus have two conditions determining the critical isotherm:

$$\left(\frac{\partial P}{\partial V}\right)_T = 0 \text{ for turning points such as B and C,}$$

$$\left(\frac{\partial^2 P}{\partial V^2}\right)_T = 0 \text{ for the coalescence of these at the point of inflexion Z.}$$

$$(6.20)$$

* A thermodynamic argument quoted in J. K. Roberts (1960), *Heat and Thermodynamics*, 5th edn, Wiley, shows that the line EAD should cut the curve at such a position that the areas DBA and ACE are equal. The reasoning is as follows. In carrying the gas through a reversible cycle DBACED (i.e. from D back to D) the gas is restored to its initial state so that there is no change in entropy. Since the whole process is at a constant temperature this implies that the net work done is zero. The work done is given by the PdV areas in each part so that the area DBA must be equal to the area ACE.

Multiplying out the van der Waals equation we have

$$PV + \frac{a}{V} - Pb - \frac{ab}{V^2} = RT.$$

Differentiating with respect to V at constant temperature we have

$$P + V\left(\frac{\partial P}{\partial V}\right) - \frac{a}{V^2} - b\left(\frac{\partial P}{\partial V}\right)_T + \frac{2ab}{V^3} = 0. \tag{6.21}$$

Inserting the condition that for all turning points $\left(\dfrac{\partial P}{\partial V}\right)_T = 0$ we obtain

$$P = \frac{a}{V^2} - \frac{2ab}{V^3}. \tag{6.22}$$

This is the equation of the curve BKZLC (figure 35). The tip of this curve corresponds to Z. This is where $\left(\dfrac{\partial P}{\partial V}\right)_T$ of equation (6.22) is zero. We have

$$\left(\frac{\partial P}{\partial V}\right)_T = -\frac{2a}{V^3} + \frac{6ab}{V^4} = 0. \tag{6.23}$$

We call the volume at which this occurs the critical volume V_c. From equation (6.23) we obtain

$$V_c = 3b. \tag{6.24}$$

Substituting in equation (6.22) we obtain for the critical pressure

$$P_c = \frac{a}{27b^2}. \tag{6.25}$$

Substituting V_c and P_c in the van der Waals equation we find the critical temperature T_c from the relation

$$\left(P_c + \frac{a}{V_c^2}\right)(V_c - b) = RT_c.$$

This gives

$$T_c = \frac{8a}{27Rb}. \tag{6.26}$$

The nature of the van der Waals equation is such that all isotherms below T_c have two turning-points; no isothermals above T_c have even a single point of inflexion. We may note that empirically the boiling point (°K.) of a liquid at atmospheric pressure is about $\frac{2}{3}T_c$. There is no simple theory for this.

6·2·7 Magnitudes of b and a in the van der Waals equation

6·2·7·1 *Meaning of b.* We have already seen that for a van der Waals gas b is four times the volume of the molecules treating them as spheres of molecular diameter σ. In a gram-molecule therefore

$$b = 4N_0 \tfrac{4}{3}\pi \left(\frac{\sigma}{2}\right)^3 = \frac{2\pi}{3} N\sigma^3. \tag{6.27}$$

From a study of the compressibility of a gas we can find the best values of b to fit the data and hence calculate σ using equation (6.27). Some typical results are given in Table 16; they are in good agreement with values of σ deduced from diffusion, viscosity, etc.

Table 16 *Values of σ deduced from b*

Gas	He	H_2	N_2	CO_2	SO_2
σ Å	2·66	2·76	3·14	3·24	3·55

6·2·7·2 *Meaning of a.* The pressure of a gas on a container is due to molecular collisions. If the velocity component normal to the container wall is u, then as we saw in Chapter 4 we may say:

$$\text{Number of collisions cm.}^{-2} \text{ sec.}^{-1} = \frac{nu}{2}$$

$$\text{Momentum change collision}^{-1} = 2mu. \tag{6.28}$$

The rate of change of momentum cm.$^{-2}$ sec.$^{-1}$ which is the product of these is the ideal gas pressure

$$P_{\text{ideal}} = nmu^2 = \tfrac{1}{3}nmc^2. \tag{6.29}$$

In escaping from the attractive force of the last few neighbouring molecules the collision velocity is reduced from u to $u - \Delta u$, so that the momentum transfer is reduced to $2m(u - \Delta u)$. The number of collisions cm.$^{-2}$ sec.$^{-1}$ which we may call the 'flux' is still $nu/2$. At first sight it might seem that this should also be diminished to $\frac{2}{n}(u - \Delta u)$. This however is not so. Although the molecules are retarded they are not, *in this model*, prevented from reaching the wall, so continuity in flux from the bulk of the gas to the wall is maintained. The flux does, of course, depend on the molecular velocity but near the wall where retardation occurs there is a slight 'crowding-up' of molecules, that is, a compensating increase in density of the gas. In this way flux-continuity is maintained. This model should be contrasted with the Dieterici treatment described above. The observed pressure P therefore becomes

$$P = \frac{nu}{2} 2m(u - \Delta u) = nmu^2 \left(1 - \frac{\Delta u}{u}\right)$$

or $\qquad P = P_i \left(1 - \frac{\Delta u}{u}\right).$ $\hfill (6.30)$

Clearly $P_{\text{ideal}} \dfrac{\Delta u}{u}$ is the van der Waals term $\dfrac{a}{V^2}$. We may write this as

$$P_{\text{ideal}} \frac{\Delta u}{u} = nmu^2 \frac{\Delta u}{u}$$

$$= \Delta(\tfrac{1}{2}mu^2)n.$$

Now $\Delta(\tfrac{1}{2}mu^2)$ is the loss in K.E. per molecule in the u direction and $n = \dfrac{N}{V}$.

Consequently the van der Waals term

$$\frac{a}{V^2} = \frac{N_0}{V} \text{ (loss of K.E. per molecule)} \hfill (6.31)$$

or $\qquad a = N_0 V \text{ (loss of K.E. per molecule)} \hfill (6.32)$

assuming no changes occur on the average in the v or w directions.

We could now assume some law of force between the molecules and integrate for all molecules and all distances, extending from the position of the last collision near the wall to all the molecules throughout the bulk of the gas. Indeed the van der Waals treatment is basically concerned with relatively long-range forces. However, as we are interested only in an order of magnitude calculation we can arrive at a reasonable value of a by considering only the nearest neighbours. Because of the simple way this is done it actually over-estimates the effect of the attractive forces. The method is illustrated in figure 36. Every molecule within the hemispherical shell bounded by $r_1 = \sigma/2$ and $r_2 = 3\sigma/2$ is a neighbour which actually touches the molecule considered.

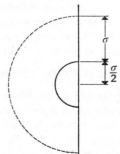

Figure 36. The range of attraction of surrounding gas molecules as a single gas molecule collides with the wall.

The work done in dragging the molecule away from these will be this number of neighbours multiplied by $\Delta\varepsilon$. The volume of the shell is $\frac{1}{2}\frac{4}{3}\pi(r_2^3 - r_1^3) = \frac{\pi}{6}13\sigma^3 = 13v_m$ where v_m is the theoretical volume of a single molecule. The number of neighbours is then $n \times 13v_m$ so that the energy lost in pulling away the molecule, i.e. its loss of K.E. in the u direction is

$$(\tfrac{1}{2}mu^2) = n \times 13 \times v_m(\Delta\varepsilon).$$

The van der Waals term is then

$$a = N_0 V n 13 v_m(\Delta\varepsilon). \tag{6.33}$$

Since $\quad b = 4N_0v_m$ and $N = Vn$,

we have $\quad a = \dfrac{13}{4}N_0b(\Delta\varepsilon).$ $\tag{6.34}$

Inserting this in our value for T_c,

$$T_c = \frac{8a}{27Rb} = \frac{8}{27}\frac{13}{4}N_0b(\Delta\varepsilon)\frac{1}{Rb} = \frac{26}{27}N_0\frac{\Delta\varepsilon}{R} = \frac{26}{27}\frac{\Delta\varepsilon}{k}$$

i.e. $\quad kT_c \simeq \Delta\varepsilon.$ $\tag{6.35}$

This extremely crude model gives a reasonable value for a and explains why the critical temperature is nearly equal to $\Delta\varepsilon/k$. It must be emphasized that this is an 'order of magnitude' calculation and that the final constant in front of $\Delta\varepsilon$ ($\frac{26}{27} \simeq 1$), is largely fortuitous. The model itself, however, is basically valid.

6·3 Some properties of the critical point

6·3·1 Critical volume

$V_c = 3b = 12 \times$ volume of molecules themselves. Hence the mean separation between the molecules at the critical point is

$$1 = (12)^{\frac{1}{3}}\sigma \simeq 2\cdot4\sigma. \tag{6.36}$$

There is just about enough room between molecules to squeeze another.

6·3·2 Mean free path

Using our simple derivation,

$$\lambda = \frac{1}{\pi n\sigma^2},$$

$$\frac{\lambda}{\sigma} = \frac{1}{\pi n\sigma^3} = \frac{1}{6nv_m}.$$

113 Imperfect gases

But nv_m is the volume occupied by the molecules in 1 c.c. of space. At the critical point this is $\frac{1}{12}$ c.c. Hence

$$\frac{\lambda}{\sigma} = \frac{1}{6 \times \frac{1}{12}}$$

$$= 2. \tag{6.37}$$

Thus σ, λ and the mean molecular separation are all roughly equal at the critical point. For ideal gases at ordinary temperatures and pressures the relevant quantities are roughly 3, 1000, 30 Å. respectively.

6·3·3 Critical coefficient

The ratio $\dfrac{RT_c}{P_c V_c}$ is known as the critical coefficient. For all equations of state which involve only two arbitrary constants the critical coefficient is an explicit dimensionless number. The values obtained for the Berthelot, Dieterici and van der Waals equations are summarized in Table 17.

Table 17 Critical points and critical coefficients

Equation of state	V_c	P_c	T_c	$\dfrac{RT_c}{P_c V_c}$
Berthelot see equation (6.15)	$3b$	$\dfrac{RT_c}{2b} - \dfrac{a}{9T_c b^2}$	$\dfrac{8a}{27Rb}$	$\frac{8}{3} \simeq 2\cdot7$
Dieterici see equation (6.16)	$2b$	$\dfrac{a}{4e^2 b^2}$	$\dfrac{a}{4Rb}$	$\dfrac{e^2}{2} \simeq 3\cdot7$
Van der Waals see equation (6.13)	$3b$	$\dfrac{a}{27b^2}$	$\dfrac{8a}{27Rb}$	$\frac{8}{3} \simeq 2\cdot7$

Experimental values for the critical coefficient are as follows:

Gas	He	H_2	N_2	CO_2	Water	Acetic acid
$\dfrac{RT_c}{P_c V_c}$	3·1	3·0	3·4	3·5	4·5	5·0

We see that, in general, for a simple gas $\dfrac{RT_c}{P_c V_c}$ lies between 3 and 3·5 if the attractive forces are of a van der Waals nature. For this particular parameter the Dieterici equation gives better agreement than the van der Waals equation. For larger and more polar molecules the critical coefficent is generally greater than 4·5.

6·4 Law of corresponding states

An interesting result is obtained if we express volumes, pressures and temperatures in terms of their critical values. We put

$$\theta = \frac{T}{T_c}, \quad \pi = \frac{P}{P_c}, \quad \phi = \frac{V}{V_c}. \tag{6.38}$$

Then the van der Waals equation becomes

$$\left(\pi + \frac{3}{\phi^2}\right)(3\phi - 1) = 8\theta. \tag{6.39}$$

We note that the van der Waals constants a and b have disappeared in this reduced equation. This is, at first sight, surprising. We know that in general we need 5 parameters to describe the behaviour of a real gas, e.g.

$$P = f(T, V, a, b).$$

However in using reduced parameters we are really feeding in two extra pieces of information $\left(\dfrac{\partial V}{\partial P}\right)_T = 0$ and $\left(\dfrac{\partial^2 V}{\partial P^2}\right)_T = 0$, on the critical isotherm, so that two of the parameters are eliminated and we are left with

$$\pi = f(\theta, \phi).$$

This, in turn, implies that if two of the quantities, say θ, ϕ, are the same for two substances the third parameter π will also be the same for the two substances. The assumption that the reduced equation is equally applicable to all gases is known as 'the law of corresponding states'. It implies that by a simple extension or contraction of the scales on which P and V are plotted all gases will give identical isotherms at the appropriate corresponding temperature. This conclusion holds quite well for gases of a single species, for example, monatomic or homonuclear diatomic gases, but for larger, more complicated molecules the agreement is poorer.

It may be noted that any equation of state which involves only two arbitrary constants can be converted into a reduced equation of state containing no arbitrary constants. For example the Dieterici equation becomes

$$\pi[2\phi - 1] = \theta \exp\left[2\left(1 - \frac{1}{\theta\phi}\right)\right]. \tag{6.40}$$

115 Imperfect gases

6·5 Expansion of gases

6·5·1 Reversible adiabatic expansion of a real gas

For an ideal gas adiabatic expansion gives a temperature drop given by

$$\frac{T_2}{T_1} = \left(\frac{V_1}{V_2}\right)^{\gamma-1}. \tag{6.41}$$

For a real gas the temperature drop is greater because of the energy expended in overcoming intermolecular attraction.

6·5·2 Joule expansion into a vacuum

For an ideal gas the internal energy is independent of the volume so that there is no net temperature change. For a real gas work is done solely against the intermolecular attractions. We may consider this as work done against the internal pressure a/V^2, and write it as

$$W = \int_{V_2}^{V_1} \frac{a}{V^2}\,dV = a\left[\frac{1}{V_1} - \frac{1}{V_2}\right], \tag{6.42}$$

where V_1 is the initial and V_2 the final volume.

6·5·3 Joule–Kelvin expansion

This is a type of expansion suggested by Lord Kelvin and carried out by him in collaboration with Joule in 1853. The purpose was to study (under more

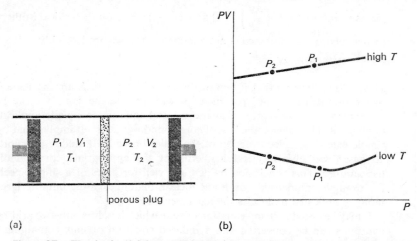

(a)   (b)

Figure 37. The Joule–Kelvin expansion. (a) The gas is driven at constant pressure P_1 through a porous plug and withdrawn at a lower pressure P_2, (b) isotherms showing that for an expansion from a higher to a lower pressure, the change in PV is positive at low temperatures and negative at high temperatures.

116   Gases, liquids and solids

carefully controlled conditions than Joule's earlier experiments) the heat changes occurring when a gas undergoes free expansion. The general principle is illustrated in figure 37(a). Gas in the left-hand part of the tube is driven at a constant pressure P_1 through an orifice or porous plug and withdrawn on the right-hand part at a constant, lower, pressure P_2. Joule and Kelvin studied the temperature changes occurring after a steady state had been reached for a system that was thermally insulated. For the purpose of our treatment, however, we shall assume that heat sources or sinks can be applied so that the temperature T on both sides is maintained constant.

Let V_1 be the volume per gram-molecule on the left-hand side, V_2 the volume on the right-hand side. During the passage of 1 gram-molecule of gas

External work done on the gas on l.h.s. $= P_1 V_1$

External work done by the gas on r.h.s. $= P_2 V_2$

Hence net work done by the gas $= P_2 V_2 - P_1 V_1.$ (6.43)

There is also work done by the gas against the internal forces. We have already calculated this in equation (6.42). The amount is

$$\frac{a}{V_1} - \frac{a}{V_2}.$$ (6.44)

Hence the total work done by the gas when one gram-molecule flows is

$$P_2 V_2 - P_1 V_1 + \frac{a}{V_1} - \frac{a}{V_2}$$

$$= \left(P_2 V_2 - \frac{a}{V_2} \right) - \left(P_1 V_1 - \frac{a}{V_1} \right).$$ (6.45)

If there is to be no cooling of the gas, this amount of heat has to be supplied. We may calculate the amount of heat involved assuming the gas obeys the van der Waals equation. We start with

$$\left(P + \frac{a}{V^2} \right)(V - b) = RT,$$

and multiply out. Neglecting the term in $\dfrac{ab}{V^2}$ we have

$$PV = RT - \frac{a}{V} + bP,$$ (6.45a)

or $$PV - \frac{a}{V} = RT - \frac{2a}{V} + bP.$$ (6.46)

In the second term on the r.h.s. we replace V by $\dfrac{RT}{P}$, so that

$$PV - \frac{a}{V} = RT - P\frac{2a}{RT} + bP.$$ (6.47)

117   Imperfect gases

J

If T is to be maintained constant

$$\left(PV-\frac{a}{V}\right)_2-\left(PV-\frac{a}{V}\right)_1=(P_1-P_2)\left(\frac{2a}{RT}-b\right).\qquad(6.48)$$

Since $P_1 > P_2$ the heat supplied to maintain T constant must be positive if $\frac{2a}{RT}-b$ is positive, i.e. if $\frac{2a}{RT} > b$. This means that if heat is *not* supplied there will be *cooling*.

Experiments show that with nearly all gases there is a cooling at room temperature after a Joule–Kelvin expansion. With hydrogen, however, there is a rise in temperature unless the gas is first cooled to about $-80°C$.

If the system is thermally isolated we may estimate the temperature difference produced during a Joule–Kelvin expansion in the following way. We write equation (6.47) more fully as

$$\left(PV-\frac{a}{V}\right)_2-\left(PV-\frac{a}{V}\right)_1=RT_2-RT_1+(P_1-P_2)\left(\frac{2a}{RT}-b\right).\quad(6.49)$$

We cannot complete this rigorously without the use of the second law of thermodynamics. It turns out that the l.h.s., which is the net work done by the gas, is equal to $C_V\Delta T$, where ΔT is the temperature drop T_1-T_2. Equation (6.49) becomes

$$C_V\Delta T= -R\Delta T+(P_1-P_2)\left(\frac{2a}{RT}-b\right).$$

Since $C_V = C_P-R$ we have finally for the cooling

$$\Delta T=\frac{1}{C_P}(P_1-P_2)\left\{\frac{2a}{RT}-b\right\}.\qquad(6.50)$$

We see that there is a critical temperature below which there is a cooling, above which a heating of the gas. This inversion temperature occurs when

$$\frac{2a}{RT_i}=b$$

or

$$T_i=\frac{2a}{Rb}=2T_B.\qquad(6.51)$$

6·5·4 *The reason for an inversion temperature*

It is instructive to consider separately the two main terms involved in the cooling of a gas subjected to a Joule–Kelvin expansion. If the gas expands from P_1 to P_2, i.e. if P_1 is greater than P_2 the work done by the gas which may cause cooling is

$$P_2V_2 - P_1V_1 + \left(\frac{a}{V_1} - \frac{a}{V_2}\right)$$

$$= \Delta[PV]_1^2 + \text{work against internal attraction.} \qquad (6.52)$$

For a real gas in which there is molecular attraction the term on the right must be positive and must therefore always lead to cooling. This is not necessarily true of the first term. At low temperatures, in moving from P_1 to P_2, the quantity $\Delta[PV]_1^2$, is positive [see figure 37(b)]. Consequently additional work must be done by the gas in expanding and this causes increased cooling. At high temperatures, however, well above the Boyle isotherm, $\Delta[PV]_1^2$ is negative, i.e. work is done on the gas as it expands and this opposes the cooling effect of the second term in equation (6.52). At some intermediate temperature the two effects just balance and there is neither heating nor cooling. This is the inversion temperature T_i.

Although this is helpful in showing how the inversion temperature arises it is, in some ways, artificial since it attempts to separate the 'imperfect' properties of the gas into two independent characteristics. As the van der Waals equation shows these two characteristics are, in fact, linked.

6·5·5 An isoenthalpic expansion in the Joule–Kelvin experiment

When a gas is expanded reversibly through nozzles or a porous plug under conditions of thermal insulation we have by the first law of thermodynamics

$$dQ = dU + PdV$$
$$0 = U_2 - U_1 + P_2V_2 - P_1V_1,$$

so that

$$U_2 + P_2V_2 = U_1 + P_1V_1$$
$$H_2 = H_1.$$

The porous-plug expansion is thus an isoenthalpic process. The temperature change associated with the pressure change must therefore be expressed in the form $\left(\dfrac{\partial T}{\partial P}\right)_H$. For a small change in H we have

$$dH = \left(\frac{\partial H}{\partial T}\right)_P dT + \left(\frac{\partial H}{\partial P}\right)_T dP. \qquad (6.53)$$

If we are specifying that H is constant $dH = 0$ and we are left with

$$\left(\frac{\partial H}{\partial T}\right)_P + \left(\frac{\partial H}{\partial P}\right)_T \left(\frac{\partial P}{\partial T}\right)_H = 0.$$

The first term on the left is C_P. Consequently

$$\left(\frac{\partial T}{\partial P}\right)_H = -\frac{1}{C_P}\left(\frac{\partial H}{\partial P}\right)_T. \qquad (6.54)$$

In the next step we have to make use of thermodynamic functions based on the second law, viz:

$$\left(\frac{\partial H}{\partial P}\right)_T - V = -T\left(\frac{\partial V}{\partial T}\right)_P \tag{6.55}$$

The Joule–Kelvin temperature change is therefore given by

$$\left(\frac{\partial T}{\partial P}\right)_H = -\frac{1}{C_P}\left(\frac{\partial H}{\partial P}\right)_T = \frac{T\left(\frac{\partial V}{\partial T}\right)_P - V}{C_P},$$

or

$$\Delta T = \frac{1}{C_P}\Delta P\left[T\left(\frac{\partial V}{\partial T}\right)_P - V\right]. \tag{6.56}$$

Using equation (6.45a) a little manipulation shows that the term $T\left(\frac{\partial V}{\partial T}\right)_P - V$ in equation (6.56) is equal to $\frac{2a}{RT} - b$. Thus equation (6.56) becomes identical with the relation given in equation (6.50) for the Joule–Kelvin cooling. Consequently the inversion temperature is again given by

$$T_i = \frac{2a}{Rb}.$$

The above treatment suggests that there is a single inversion temperature which is independent of pressure. This results from assuming that the pressures are low, so that $\frac{ab}{V^2}$ can be neglected. If the higher terms are included it may be shown that for the van der Waals equation of state, T_i is a parabolic function of the pressure – indeed at pressures above $\frac{a}{3b^2}$ the gas expansion shows inversion at all temperatures, that is at high pressures it will heat up on expanding.

This is illustrated schematically in figure 38 (taken from F. W. Sears (1953), *Thermodynamics, The Kinetic Theory of Gases, and Statistical Mechanics*, 2nd edn, Addison–Wesley) where curves of constant enthalpy have been plotted as a function of temperature and pressure. In expanding from say point D to point E the temperature of the gas will rise – whereas an expansion from point A or point B to point C will produce a drop in temperature.

In effect the temperature T_i given in equation (6.51) for a low pressure expansion is the highest value of the inversion temperature. Some approximate values of T_i at low pressures are given in Table 18 and compared with $2T_B$.

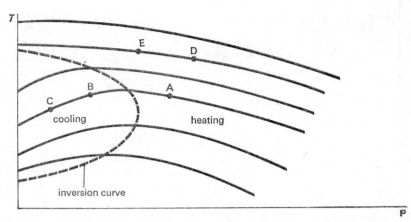

Figure 38. Curves of constant enthalpy as a function of temperature and pressure (from F. W. Sears (1953), *Thermodynamics, The Kinetic Theory of Gases, and Statistical Mechanics*, 2nd edn, Addison–Wesley).

Table 18 *Inversion temperatures T_i of some gases compared with $2T_B$ ($°K$.)*

Gas	T_B	$2T_B$	T_i
He	16	32	25
H_2	114	228	190
N_2	420	840	650

Chapter 7
The solid state

We pass at once from gases to solids. In gases the atoms and/or molecules are almost completely free; in solids they are almost completely lacking in mobility. Indeed in a solid the thermal motion is so greatly reduced that individual atoms can move from their fixed positions only with the greatest difficulty. It is this which imparts to solids their most characteristic macroscopic property – they maintain whatever shape they are given, they have appreciable stiffness. Although the atoms are in fixed locations they possess some thermal energy – in some cases individual atoms can diffuse through the solid but in general the process is extremely slow. Their main thermal exercise consists in vibrating about an equilibrium position.

7·1 Types of solids

There are three main types of solids:

7·1·1 *Crystalline solids*

The main feature here is long-range order. The molecules or atoms are in regular array over extended regions within each individual grain or crystal. With large single crystals the regular array may extend over an enormous number of atoms or molecules. For example with a single crystal of copper of side 1 cm., along any one direction there will be 30,000,000 copper atoms in regular array, the arrangement of the last few being (in the ideal case) in perfect step with the first.

In spite of this extraordinary degree of regularity the individual atoms can vibrate over relatively large amplitudes about their equilibrium positions. Thermal vibrations can easily exceed one tenth of the mean spacing; yet the mean spacing is precise and determinate to an extremely high degree of reproducibility.

7·1·2 *Glasses*

Here the individual atoms or molecules are in ordered array over a short range but there is no long-range order. They are indeed, in structure, very much like an instantaneous snap-shot of a liquid and are sometimes called supercooled liquids. However molecular movement, a characteristic feature of liquids, is minute and the attempt to describe their flow properties in terms of an equivalent viscosity leads to astronomical values of the viscosity. For example from the viscosity of certain silicate glasses at elevated temperatures it is possible to extrapolate and deduce a room temperature value of the order of 10^{70} poise.

An American physicist has pointed out that this is equivalent to stating that if the glass were in the form of a rod of length equal to the circumference of the Universe and was subjected to half its breaking tension for a period of time equal to the age of the Universe, its total extension could amount to less than one thousandth of the diameter of an electron. [F. W. Preston (1942), *Journal of Applied Physics*, vol. 13, p. 623.] For such materials, room temperature viscosity is clearly not a very useful physical concept.

Glasses may also be described as amorphous solids.

7·1·3   *Rubber or polymers*

These consist of long organic molecules held together either by chemical cross links, by hydrogen bonds or simply by van der Waals forces. Below a critical temperature (the glass transition temperature) they behave very much like glasses, but with a greater degree of ductility; above this temperature they are generally rubber-like (see section 8·4 below).

7·2   Main types of bonding in crystalline solids

We summarize here four main types of bonding and the corresponding crystalline states.

Bonding	Bond strength	Examples	Crystalline state
Van der Waals	Weak	Solid H_2, Kr, paraffins	Close packing of weakly attracted units.
Ionic	Strong	NaCl crystal	Giant aggregates of positive and negative ions clearly packed in a way consistent with neutrality of charge.
Covalent	Strong	Diamond, Si, Ge	Giant molecules with directed bonds, packing determined by valency number and valency directions.

Metallic	Strong	Metals	Metal atoms give up their valency electrons leaving metal ions in a sea of electrons. Forces between ions and electrons are central giving close packing. Strong attraction gives strength. Mobile electrons give conductivity.

7·3 Solid–liquid transitions

If we heat a solid, the molecules or atoms acquire sufficient energy to escape from their lattice sites and ultimately melt, the temperature remaining constant until the whole of the specimen is molten. The quantity of heat required to convert a unit mass of the solid at its melting point to a liquid is called the latent heat of fusion.

The melting point of a solid depends on the pressure to which it is subjected. For a solid which is more dense than the liquid (this is the usual case) the melting point increases with pressure: for a solid which is less dense (e.g. ice) the melting point decreases with pressure [see figure 39(a) and (b)]. If we

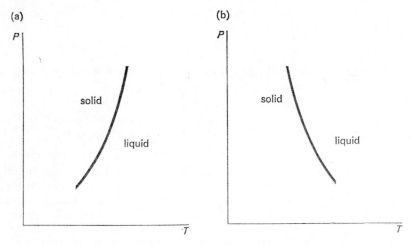

Figure 39. Melting curves (solid–liquid equilibrium) as a function of pressure for (a) a solid more dense than the liquid, (b) a solid less dense than the liquid.

combine this with the liquid–vapour transition we obtain a phase diagram as shown in figure 40. We note the following characteristics:

124 Gases, liquids and solids

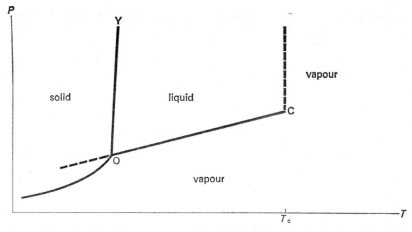

Figure 40. Equilibrium curves for solid, liquid and vapour. O is the triple point where solid, liquid and vapour are in equilibrium. The liquid phase does not persist above the critical temperature.

(a) for the liquid–vapour phase equilibrium the curve ends at C. Above this point which corresponds to the critical temperature T_c the liquid ceases to exist;

(b) by contrast for the solid–liquid phases there is no evidence that there is an end point to OY above which only one phase can exist.

7·4 Consequences of interatomic forces in solids

We describe here a few of the properties of solids which are directly explicable in terms of interatomic forces.

7·4·1 *Heat of sublimation*

If we could take a van der Waals solid such as solid argon at absolute zero, and heat it until all the constituent atoms evaporate we should have pumped into the solid sufficient energy to break the bonds between all the atoms. This is the theoretical heat of sublimation L_S. It is equal to the latent heat of melting (L_F) plus the latent heat of vaporization (L_V) plus a specific heat term involved in raising the temperature from absolute zero to the temperature of vaporization. If there were an exact energy balance the released atoms would just leave the solid with a minute escape velocity; in practice they will have considerable thermal energy. Thus precise comparisons between bond strengths and energy of sublimation is difficult. However as an order of magnitude calculation such correlation is easy, as we shall now show.

If the potential energy between each atom (or each molecule) and its neighbour is $\Delta\varepsilon$, and if each atom (or molecule) has n nearest neighbours,

125   The solid state

the energy required to break the bonds in a gram-atom (or molecule) will be

$$L_S = \tfrac{1}{2}N_0 n\Delta\varepsilon.$$ (7.1)

This is allowing only for nearest-neighbour interactions, where N_0 is Avogadro's number and the factor $\tfrac{1}{2}$ is introduced to avoid counting each bond twice. For close packing, such as in a face-centred cubic structure or a hexagonal close-packed structure, n is 12.

Hence

$$L_S = 6\Delta\varepsilon N_0.$$ (7.2)

If $\Delta\varepsilon$ is in ergs and we wish to express L_S in calories per gram-atom (or molecule) we obtain

$$L_S = 86 \times 10^{15}\Delta\varepsilon.$$ (7.3)

Some typical results are given in Table 19, the 'observed' values of L_S being simply the sum of L_F and L_V. The values of $\Delta\varepsilon$ are obtained either from theoretical calculations of the van der Waals forces, or from the behaviour of the substance in the gaseous state (for example, the deviation from idea gas behaviour at high pressures).

Table 19 *Heats of sublimation*

Substance	$\Delta\varepsilon$ erg	L_S cal. g-atom^{-1} or g-molecule^{-1}	
		Calculated	Observed
He	$0\cdot9 \times 10^{-15}$	80	20
Ne	$4\cdot8 \times 10^{-15}$	400	300
Ar	$16\cdot5 \times 10^{-5}$	1400	1800
Kr	25×10^{-15}	2200	2500
H$_2$*	4×10^{-15}	350	250
N$_2$*	13×10^{-15}	1100	1300

* Not face-centred cubic structure.

Apart from helium, where the effects of 'zero-point energy' are important, the agreement is very satisfactory. A more detailed consideration of lattice energy, latent heat of sublimation and intermolecular forces will be given in the next chapter. At this stage we merely draw attention to the direct agreement with the simplest types of solids. We may also note that the latent heat of fusion is generally much lower than the latent heat of vaporization. From the point of view of *thermal energy* the liquid state, near the melting point, is much nearer to the solid state than it is to the gaseous state.

7·4·2   Surface energy

The atoms in the bulk of a solid are subjected to the attraction of atoms all round them, but at the free surface they are subjected only to the inward pull of the atoms within the solid. Consequently the surface layers have a higher energy than those in the bulk and this excess is called the free surface energy γ.

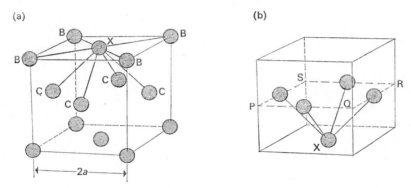

Figure 41.   Face-centred cubic crystal of side $2a$. (*a*) The surface atom X has 8 nearest neighbours at B, B, B, B and C, C, C, C; whereas in the bulk each atom has 12 nearest neighbours. (*b*) If the crystal is pulled apart in the plane PQRS each atom such as X breaks 4 bonds.

We may calculate this in terms of the bond energy for, say, a typical face-centred cubic solid where all atoms except those in the surface have 12 nearest neighbours. An atom X in the surface has only 8 nearest neighbours. This is shown in figure 41(*a*). If the crystal contains a total of N atoms and the surface contains N_s atoms the total bond energy of the crystal, counting only nearest-neighbour interactions, is then

$$\tfrac{1}{2}\Delta\varepsilon\{12(N-N_s)+8N_s\}$$
$$= \tfrac{1}{2}\Delta\varepsilon\{12N-4N_s\}. \tag{7.4}$$

(This assumes that there is no change in lattice spacing in the surface layers, a point that is still in some dispute.)

If there were no surface atoms the energy would be $\tfrac{1}{2}\Delta\varepsilon \times 12N$. Thus the energy is reduced by $\tfrac{1}{2}\Delta\varepsilon \times 4N_s$. Since $\Delta\varepsilon$ is negative this is a positive increase in energy which is indeed the surface energy. If the surface area of the crystal is A,

$$\gamma A = \tfrac{1}{2}\Delta\varepsilon \times 4N_s. \tag{7.5}$$

We may obtain this result more simply by considering a rectangular bar of crystal of cross-sectional area A. Any section across the crystal contains N_s atoms, so if we pull the crystal in two each surface atom breaks four bonds and the work done per atom is $4\Delta\varepsilon$ [see figure 41(b)]. Hence the total work done is $4\Delta\varepsilon \times N_s$. But the area exposed is $2A$.

Hence

$$2\gamma A = 4\Delta\varepsilon \times N_s. \tag{7.5a}$$

This is identical with equation (7.5). If in the crystal orientation we have considered there are z atoms per sq. cm. we see from equations (7.5) and (7.5a) that

$$\gamma = 2 \times \Delta\varepsilon \times z. \tag{7.6}$$

But the energy necessary to sublime z atoms from the bulk of the crystal [see equation (7.2)] is

$$l_s = 6\Delta\varepsilon z. \tag{7.7}$$

Hence the surface energy is equal to one third of the heat of sublimation of a single atomic layer of atoms. For different orientations the fraction will be a little different but generally it will be of the order of $\frac{1}{3}$ to $\frac{1}{2}$. For a face-centred cubic lattice of side $2a$ as shown, the volume occupied by each atom is $2a^3$. If M is the molecular weight and ρ the density

$$\frac{M}{N_0\rho} = 2a^3 \quad \text{or} \quad a^3 = \frac{M}{2N_0\rho}.$$

For the (100) face that we have considered the area occupied by each atom is $2a^2$. The number of atoms z per sq. cm. is $\dfrac{1}{2a^2}$. Hence

$$z = \frac{1}{(2a^2)} = \frac{1}{2}\left(\frac{2N_0\rho}{M}\right)^{\frac{2}{3}} \tag{7.8}$$

$$= 0 \cdot 8 \left(\frac{N_0\rho}{M}\right)^{\frac{2}{3}}.$$

Hence $\quad \gamma = \dfrac{0 \cdot 8}{3}\left(\dfrac{L_s}{N_0}\right)\left(\dfrac{N_0\rho}{M}\right)^{\frac{2}{3}}. \tag{7.9}$

This of course refers to a specific face and to a specific structure. For other faces and structures we should merely change the numerical factor by a small amount. A reasonable average value for the factor in front of equation (7.9) is $0 \cdot 3$. This relation should apply best to van der Waals solids such as solid neon, argon and krypton. However the surface energy of such materials is not known. We have therefore applied the model to metals, the results being given in Table 20. The quantitative agreement between the calculated and observed values is poor, but the agreement in relative values is surprisingly good. If all the calculated values are multiplied by a correction factor of 0·4 the calculated values all agree with observation to within about 10 per cent. The basis for this correction factor is discussed in the next chapter.

Table 20 *Surface energies of some metals in solid state*

Metal	M.P. °C.	L_s kcal. mole^{-1} at 25°C.	M g.	ρ g.cm.$^{-3}$	γ erg cm.$^{-2}$	
					Calc.	Obs.*
Indium	156	57	115	7·3	1,300	~450
Lead	327	47	207	11·3	1,200	~500
Gold	1063	90	197	19·3	3,300	1,300
Iron	1537	95	56	7·9	4,000	2,000
Platinum	1769	135	195	21·5	5,000	2,100
Tungsten	3380	203	184	19·3	6,700	2,900

* Observed measurements are generally at elevated temperatures so that they are some-what smaller than the values to be expected at 25°C. These values are from M. C. Inman and H. R. Tipler (1963), 'Interfacial energy and composition in metals and alloys', *Metallurgical Reviews*, vol. 8, 1963, pp. 105–66.

With liquids, the surface energy is relatively easy to determine; with solids, particularly ductile solids, it is far more difficult. The most direct method with metals is to hang a wire of uniform cross-section in a chamber at constant temperature and observe the region where necking just begins. (Such experiments give meaningful results only at temperatures approaching the melting point.) If necking occurs at a height h above the free end it implies that

Weight of wire of length $h = 2\pi r \gamma$,

where r is the radius of the wire. For most metals γ turns out to be of the order of 500–2000 erg cm.$^{-2}$

With a 'perfectly brittle' solid like mica the surface energy can be determined by finding the force required to open a crack. If we consider a cleaved specimen as shown in figure 42(*b*) we may allow the force F to move and to open up the crack as shown. The work done $= \int_{y_1}^{y_2} F dy$; this is expended in changing the elastic strain energy in the two mica leaves and also in creating two new surfaces each of area A. The work done can be measured, the elastic energy calculated and the difference may then be put equal to $2\gamma A$. For mica in a good vacuum γ has a value of about 5000 erg cm.$^{-2}$; in air about 300 erg cm.$^{-2}$ In general such experiments cannot be carried out satisfactorily with ductile solids such as metals or polymers, since the plastic work absorbed

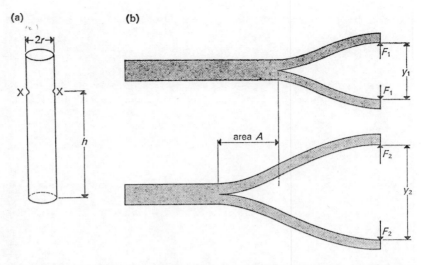

Figure 42. Determination of surface energy γ of solids (*a*) by the necking of a specimen at section XX; this occurs when the weight of the wire of length $h = 2\pi r\gamma$ (*b*) by bifurcation of a mica strip. The work done in opening up the strip is equal to the change in elastic strain energy plus the increase in surface energy $(2\gamma A)$ involved in creating 2 surfaces of area A.

by deformation at the tip of the crack may be several hundred times greater than the surface energy. For example, values of γ often come out to be of the order of 10^5 erg cm.$^{-2}$ Therefore the experiments can be carried out only on brittle solids. Reasonable values have been obtained on glass (A. A. Griffith 1920) and as mentioned above on mica (J. W. Obreimov 1931, A. I. Bailey 1955). Recently Gilman has shown that the technique can be applied successfully to many crystalline solids if the experiments are carried out at liquid air temperature – he showed that at this temperature the ductility is 'frozen out'. Surface energies of the order of a few hundred erg cm.$^{-2}$ are obtained for materials such as NaCl, LiF.

We may mention here a simple feature involved in the surface energy of solids which differentiates it markedly from that of liquids. Suppose we have a surface of area A possessing surface energy γ. If we increase the area by a small increment the work done per unit increase may be written

$$\frac{d(A\gamma)}{dA} = \gamma + A\frac{\partial\gamma}{\partial A}. \tag{7.10}$$

For liquids the term $\frac{\partial\gamma}{\partial A}$ is zero, since, on account of atomic mobility, the structure of a liquid surface is unchanged when the surface area is increased. This is not so with solids; when the surface is stretched (even only notionally) the atoms are pulled apart and γ diminishes so that $\dfrac{\partial \gamma}{\partial A}$ is negative. This implies that with solids the surface energy is not the same as the surface force (more strictly the line tension in the surface; see Chapter 10 on the surface tension of liquids); with liquids they are identical.

Although one cannot equate surface energy with surface force there is generally some tension in the free surface. Even if it is as large as an effective line-tension of 1000 dyne cm.$^{-1}$, this is minute compared with the elastic modulus of the solid, usually about 10^{12} dyne cm.$^{-2}$ Clearly forces of this magnitude can have very little visual effect on metals; so they do not influence the shape of a solid in the same way as they do liquids. A centimetre cube of copper is not dragged into spherical shape at room temperature because of surface tension. However, with very minute particles some such effects may be observed.

Surface energy of solids plays a very important part in sintering; and it is a crucial factor in the wetting of solids by liquids. From the point of view of atomic mechanisms it is worth emphasizing that the same forces which produce the pressure defect in a real gas, are responsible for the surface energy of a solid and the surface tension of a liquid.

7·4·3   *Elastic stiffness*

If atoms or molecules in a solid are subjected to a stress and thereby displaced from their equilibrium positions the interatomic forces will tend to restore them to their original positions. If the distortions are not too large, when the stress is removed the atoms will (under ideal conditions) return to exactly the same situation as before; the elastic modulus is defined as the ratio of the applied stress to the deformation produced. This will be discussed in detail in a later chapter. Here we merely quote the result for Young's modulus which is defined as

$$E = \frac{\text{Tensile stress increment } \Delta\sigma}{\dfrac{\text{increase in length } \Delta x}{\text{original length } x}}$$

$$= x \frac{\Delta\sigma}{\Delta x}. \tag{7.11}$$

The elastic modulus of a solid may be easily demonstrated if one considers the force–displacement curve. Figure 43 shows the force between two neighbouring atoms as a function of separation. At low temperatures where thermal energy can be neglected, the equilibrium separation occurs where the attractive force equals the repulsive force. Thus OA is the value of the equilibrium separation x. The slope of the line ZAY through A is $\dfrac{dF}{dx}$. Hence the quantity

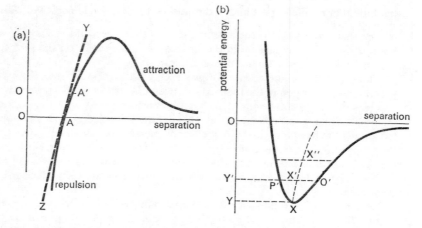

Figure 43. (*a*) Force–separation diagram illustrating the decrease in elastic modulus with expansion. (*b*) Potential energy–separation diagram illustrating the increase in separation which occurs when the temperature is increased (thermal expansion).

$x\dfrac{dF}{dx}$ is the exact one-dimensional analogue of Young's modulus as defined above.

If the solid is heated thermal expansion (see below) shifts the equilibrium position from A to A'. The slope $\dfrac{dF}{dx}$ is seen to be smaller and this decrease is appreciably greater than the corresponding increase in the new equilibrium separation O'A'. Thus the effective modulus is reduced. If we tried to construct a simple cubic structure out of these pairs of atoms the number in any square cm. of an atomic plane would be $\dfrac{1}{x^2}$ so that the stress (as distinct from the force) is $\dfrac{F}{x^2}$ and the strain is $\dfrac{dx}{x}$. Then Young's modulus becomes $\dfrac{1}{x}\dfrac{dF}{dx}$. In this case we see that the slight increase in x, with rise in temperature, augments the decrease in $\dfrac{dF}{dx}$ so that both terms of the modulus are reduced by a rise in temperature. However as already mentioned the change in x is in general small compared with the change in $\dfrac{dF}{dx}$ since F is usually a 'strong' power function of x.

We conclude that if the behaviour of the solid as a whole resembles that of the diatomic model discussed above the elastic modulus decreases as the temperature is increased. This is true of practically all solids except rubber-like solids.

7·4·4 *Thermal expansion*

If the length of a solid specimen at 0°C. is l_0 and at t°C. is l_t one can usually write

$$l_t = l_0(1 + \alpha t),$$

where α is the coefficient of linear expansion and for most solids is of order 10^{-5} °C.$^{-1}$

This behaviour may again be explained in terms of the forces between the individual atoms. It is most convenient to plot the potential energy curve as a function of separation. If we do this for two atoms we obtain the typical curve shown in figure 43(*b*). At absolute zero the equilibrium position (ignoring zero-point energy) is at X, so the equilibrium separation is YX. As the temperature is raised the energy rises to, say, Y' and one of the atoms will oscillate relative to the other between positions P' and Q' about a mean position X'. Because of asymmetry in the shape of the potential energy curve the length Y'X' is greater than YX. Thus the mean separation increases with temperature along the curve XX'X" . . . .

If an assembly of atoms in a solid behaved similarly this would explain thermal expansion. Of course there are added complications. In a solid one ought to treat this problem in terms of the free energy; this introduces an additional term but it is generally small compared with the effect described above. Again in the diatomic model the atoms dissociate when the energy reaches 0. This is equivalent to sublimation of the solid and tells us nothing of the melting process that occurs somewhere en route. This is because melting is a co-operative process involving many atoms and a satisfactory model must be multi-dimensional to allow for their behaviour (see Chapter 10). Finally we may note that increased temperature may give rise to transverse oscillations. In some types of crystals this may actually lead to a contraction in one crystallographic direction as the temperature is raised.

Nevertheless the simple model given above provides a reasonably satisfactory explanation of thermal expansion in solids. In a later chapter we shall indeed use it to derive an explicit relation for the thermal expansion in terms of intermolecular forces. We shall see that the results are in quantitative agreement with experiment. In general the potential energy–separation curve provides a very useful way of approaching many of the physical and mechanical properties of solids.

133   The solid state

Chapter 8
The elastic properties of solids

If we subject a solid rod to tension we pull the atoms further apart; if we compress it we push them closer together. The interatomic forces resist these changes and, if the distortions are not too large, the body will return to its original dimensions when the external forces are removed; the deformations are then called elastic.

8·1   Some basic elastic properties

8·1·1   *Elastic moduli*

If a solid rod of uniform cross-sectional area A is pulled with a tensile force F the tensile stress is defined as

$$\sigma = \frac{F}{A}.$$

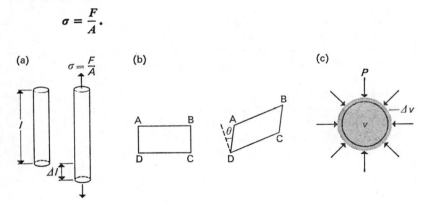

(a) $\sigma = \dfrac{F}{A}$ (b) (c)

Figure 44.   (a) Deformation in tension, (b) deformation in shear, (c) deformation under hydrostatic pressure.

If, as a result, the rod increases its length from l to $l+\Delta l$ the fractional increase $\dfrac{\Delta l}{l}$ is called the linear strain ε [see figure 44(a)]. Within the elastic range Hooke showed that the stress is proportional to the strain. The ratio is defined as Young's modulus E:

$$E = \frac{\text{Linear stress}}{\text{Linear strain}} = \frac{\dfrac{F}{A}}{\dfrac{\Delta l}{l}}. \tag{8.1}$$

A solid may also be deformed in shear. For example if we consider a rectangular element as in figure 44(b), we may apply a shear stress τ (force ÷ area) across the faces AB and CD. This must be accompanied by equal shear stresses τ along CB and AD so that the net couple is zero (see Chapter 9, section 9·12). The element is then distorted through a small angle θ which is termed the shear strain. The strain is again found to be proportional to the stress and the ratio is termed the rigidity modulus n:

$$n = \frac{\tau}{\theta}. \tag{8.2}$$

A solid may also be deformed by a hydrostatic pressure p. If the solid has volume v, and the change in volume is $-\Delta v$, then the volume strain is $-\Delta v/v$ and this is found to be proportional to p. The ratio is the bulk modulus K.

$$K = \frac{p}{-\dfrac{\Delta v}{v}}. \tag{8.3}$$

The reciprocal of this modulus is known as the compressibility, β.

Finally when a body is extended in tension to give a linear strain ε_l, it contracts in directions at right angles to the direction of tension. The fractional contraction ε_t in the transverse direction is found to be proportional to ε_l. The ratio

$$\frac{\text{Contractile strain}}{\text{Linear strain}} = \frac{\varepsilon_t}{\varepsilon_l} = \text{Poisson's ratio } v. \tag{8.4}$$

For most solids v has a value between $\frac{1}{4}$ and $\frac{1}{3}$ and this implies a decrease in volume under tension. If $v = \frac{1}{2}$ (as is nearly the case with rubber) there is no volume change in tension or compression.

There are several relationships between E, n, K and v which we shall not prove here. For example

$$n = \frac{E}{2(1+v)} \tag{8.5}$$

$$K = \frac{E}{3(1-2v)}. \tag{8.6}$$

We see from equation (8.6) that for values of $v = 0·25$ to $0·33$, K is comparable with E; for a value of $v = \frac{1}{2}$, $K = \infty$. This implies that the material is incompressible, which is what would be expected since tension and compression produce no volume change in such a material.

8·1·2 *Torsional rigidity*

If a circular tube of internal radius r_1, external radius r_2, length l is held at one end and subjected to couple Γ at the other end, every part of the tube is subjected to shear. Suppose the free end is twisted through an angle ϕ. The ratio Γ/ϕ is known as the torsional rigidity. We shall not derive it but simply quote the relationship of Γ/ϕ to be:

$$\frac{\Gamma}{\phi} = \frac{n\pi}{2l}\,[r_2^4 - r_1^4], \tag{8.7}$$

where n is the rigidity modulus.

8·1·3 *The bending of a cantilever*

If a bar is encastered at one end and deflected sideways by a force F at the other end, the deflection produced depends on F, on the section of the bar, its length and its Young's modulus. Although we shall not derive the equation here, we shall indicate the principle of the analysis involved. Suppose figure 45(*a*) represents, in exaggerated form, the bending of a beam. The sections p, q, r, s, t indicate how the beam is deformed. We make a notional cut through the bar at section r; then the material to the left of r is subjected to a bending moment Fx. If the beam is to be in equilibrium this must be counter-balanced by a reverse moment of the same magnitude. The material above AB is stretched and so exerts a tensile force to the left. The material below AB is compressed and so exerts a resisting force to the right. The beam bends to just that point where these two forces produce a couple exactly equal to Fx. This implies that between the stretched and compressed portions there is a line AB which is unchanged in length. AB is, in fact, a part of the line XX′ which represents the original length of the unstressed bar; it is known as the neutral axis. We may note in passing that the section between s and t has to carry the smallest bending moment and is therefore distorted least: the end part of the beam is indeed straight. On the other hand the bending moment is a maximum at CD; here the curvature is a maximum. This raises an interesting point which is usually ignored by physicists and engineers. The encastered portion CDEF is considered completely rigid and subjected to zero bending moment. This implies that at the boundary CD the bending moment suddenly drops from a maximum value on the right to zero on the left. Such a situation is impossible, for there must be a transitional region HK within the encastered zone where the bending moment gradually falls to zero. In practice this implies a certain amount of slip between the bar and the clamp at ED and FC or some other type of yielding in the clamped zone. This is not of general importance in calculating the deflection of a loaded bar. It can however be of very great importance; for example in the propagation of a crack or in the splitting of mica described earlier as a means of deter-

136 Gases, liquids and solids

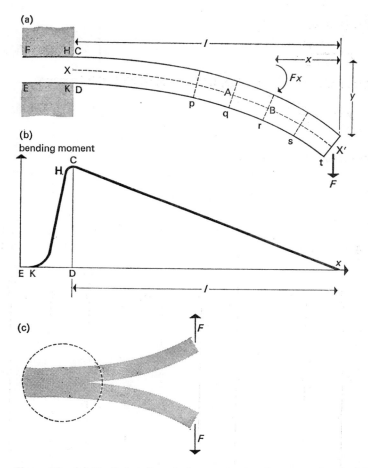

Figure 45. (*a*) Elastic bending of a beam encastered at one end and subjected to a force *F* at the other. The line XX' is the neutral axis where there is neither extension nor compression of the original length. (*b*) The bending moment has a minimum at the free end and a maximum at the point where the beam enters the encastered zone. Simple theory assumes that it then drops at once to zero; in practice there must be some transition region HK. (*c*) The bifurcation region of a mica specimen resembles the encastered zone of a bent beam.

mining the surface energy. The conditions at the region of bifurcation resemble those of the encastered region in a bent beam. The analysis of the forces in this region is very difficult indeed.

If the moment of inertia of the section of the beam about the vertical axis is *I* and the length of the beam is *l* the deflection *y* produced at the end by a

137   The elastic properties of solids

force F is given by

$$y = \frac{F}{EI} \frac{l^3}{3}.$$ (8.8)

For a beam of rectangular section, thickness a, width b, this becomes

$$y = \frac{F}{E} \times \frac{4l^3}{ab^3}.$$ (8.9)

8·1·4 *Young's modulus in terms of interatomic force constant*

In the previous chapter we showed how the elastic modulus is related to the force displacement curve. We shall now make this more quantitative. Consider a solid composed of a single species of atom and let us assume a simple cubic structure, the separation between each atom being a. Let the bar have unit cross-sectional area and consider three neighbouring planes I, II and III (figure 46). We assume that there are only nearest-neighbour interactions;

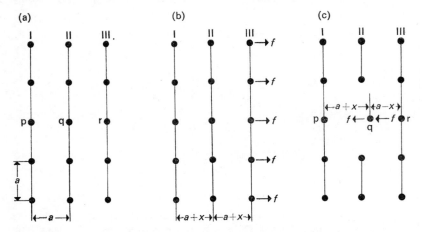

Figure 46. Simple cubic array of atoms, (a) undeformed, (b) when a force f is applied to each atom in plane III all the planes undergo an increased separation x, (c) a single force f acts in a similar way on a single atom q.

then if an atom q in plane II is displaced in a direction normal to the plane, it experiences a restoring force only from atom p in plane I and atom r in plane III. If a tensile force f is applied to atom r the spacing between the atoms in that row will each be increased by a small distance x (where $x \ll a$). The restoring force which each atom in this row experiences will be proportional to x. Let the restoring force be kx where k is, in fact, a measure of the slope of the force–displacement curve shown in the previous chapter [figure 43(a)].

Since there are $1/a^2$ atoms per sq. cm., the force per sq. cm. (the stress) to achieve a separation of the planes of this amount will be $\dfrac{f}{a^2} = \dfrac{kx}{a^2}$. This displacement is equivalent to a linear strain of amount $\dfrac{x}{a}$. Then we may write for Young's modulus

$$E = \frac{\text{Tensile stress}}{\text{Strain}} = \frac{\dfrac{kx}{a^2}}{\dfrac{x}{a}} = \frac{k}{a}. \tag{8.10}$$

8·2 Propagation of longitudinal waves along an elastic bar

We extend this to a very simple description of wave propagation. Our major (and weakest) assumption is that we may isolate the interaction of one atom with its next neighbour. Suppose we hold planes I and III fixed [see figure 46(c)] and displace atom q relative to atoms p and r by an amount x. Then p exerts an attractive force kx on q and r exerts a repulsive force of the same magnitude. The resulting restoring force is $2kx$. If the atomic mass is m the equation of motion is

$$m\ddot{x} + 2kx = 0.$$

This gives simple harmonic motion of period

$$T = 2\pi \sqrt{\frac{m}{2k}}. \tag{8.11}$$

Suppose now we give an impulse to the free end of the bar and displace the row of atoms in plane I to the right (figure 46). These then act on the atoms in row II which in turn are pushed to the right and so act on the atoms in plane III. In this way the impulse is propagated through the solid. The time taken for each plane to respond to the push of the neighbouring plane is of the order of one quarter of the vibrational period T. Thus the impulse jumps from one plane to the next in a time

$$t \simeq \frac{2\pi}{4} \sqrt{\frac{m}{2k}} \simeq \frac{2\pi}{5\cdot7} \sqrt{\frac{m}{k}}. \tag{8.12}$$

For our purpose we may take $2\pi/5\cdot7$ to be unity. The disturbance thus travels a distance a in time $\sqrt{\dfrac{m}{k}}$. The velocity of propagation is therefore

$$v = \frac{a}{t} = a \sqrt{\frac{k}{m}},$$

or

$$v = \sqrt{\left(\frac{a^3}{m}\frac{k}{a}\right)}. \tag{8.13}$$

139 The elastic properties of solids

We recognize $\frac{m}{a^3}$ as the density ρ, and the quantity $\frac{k}{a}$ as Young's modulus E. Hence

$$v = \sqrt{\frac{E}{\rho}}. \tag{8.14}$$

This is the same as the equation for the velocity of a longitudinal wave in a bar in terms of its macroscopic properties. It shows that the vibrating atom is the 'messenger' for wave propagation.

The frequency of vibration is

$$v = \frac{1}{T} = \frac{1}{2\pi}\sqrt{\left(\frac{2k}{m}\right)} = \frac{\sqrt{2}}{2\pi a}v,$$

$$v = \frac{1}{2\pi a}\left(\frac{2E}{\rho}\right)^{\frac{1}{2}}. \tag{8.15}$$

For, say, iron we have $E = 2 \times 10^{12}$ dynes cm.$^{-2}$, $\rho = 7 \cdot 8$ g. cm.$^{-3}$ and $a \simeq 2 \cdot 5 \times 10^{-8}$ cm. Then $v \simeq 5 \times 10^{12}$ vibrations per sec. This is the right order of magnitude.

The main defect in this model is that the vibrations of the atoms are not isolated but are coupled. A rigorous treatment does not, however, substantially alter the nature of this analysis.

8·3   Bulk moduli

8·3·1   *Bulk modulus of an ionic solid*

We consider here an ionic solid of the NaCl type in which the Na^+ and Cl^- ions are distributed on a cubic lattice as shown in figure 47. We consider a single ion and its coulombic interaction with all the charged ions around it. As the figure shows there are:

> 6 neighbours of opposite sign at distance x
> 12 neighbours of same sign at distance $\sqrt{2}x$
> 8 neighbours of opposite sign at distance $\sqrt{3}x$, etc.

If each ion carries charge e (or $-e$) the potential energy of one ion in relation to all its neighbours is

$$\frac{-e^2}{x}\left[6 - \frac{12}{\sqrt{2}} + \frac{8}{\sqrt{3}} \cdots\right]$$

$$= \frac{-e^2}{x}[2 \cdot 1 \ldots]. \tag{8.16}$$

Proper summation to infinity shows that the correct value, known as the Madelung constant A, is $1 \cdot 75$ for the rock-salt structure.

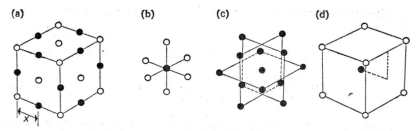

Figure 47. The structure of the NaCl crystal. (a) General view. (b) 6 neighbours of opposite sign at distance x. (c) 12 neighbours of same sign at distance $\sqrt{2}x$. (d) 8 neighbours of opposite sign at distance $\sqrt{3}x$.

Assuming a repulsive potential energy term between each ion pair of the form A/x^9, the summation of all the repulsion terms will give B/x^9 where B is an appropriate constant. The potential energy of one ion allowing for both attraction and repulsion is

$$u = -1.75\frac{e^2}{x} + \frac{B}{x^9}.$$ (8.17)

If there are N positive and N negative ions per mole the potential energy for a mole of NaCl is

$$U = \tfrac{1}{2}(2N_0 u) = N_0\left(-1.75\frac{e^2}{x} + \frac{B}{x^9}\right)$$ (8.18)

where the $\tfrac{1}{2}$ is introduced so that we do not count each bond twice.

The equilibrium spacing a occurs when U is a minimum.

Putting $\dfrac{\partial U}{\partial x}$ at $x = a$ equal to zero we find that B has the value $\dfrac{1.75e^2 a^8}{9}$.

Equation (8.18) becomes

$$U = -1.75 N_0 e^2 \left[\frac{1}{x} - \frac{a^8}{9x^9}\right].$$ (8.19)

The next step appears complicated but is basically an attempt to connect the internal energy with an elastic modulus, in this case the bulk modulus K. If we apply a hydrostatic pressure p this produces a volume change $-dV$, and the work done by the external forces on the crystal is then $-pdV$. The ions are squeezed closer together and if the whole of this work goes solely in increasing that part of U which arises from interatomic forces, we have $-pdV = dU$ or

$$p = -\frac{\partial U}{\partial V}.$$ (8.20)

But $$K = -\frac{dp}{\dfrac{dV}{V}} = -\frac{\partial^2 U}{\partial V^2}V.$$ (8.21)

We need to express this in terms of the ionic spacing x. We have

$$\frac{\partial^2 U}{\partial V^2} = \frac{\partial}{\partial V}\left(\frac{\partial U}{\partial V}\right) = \frac{\partial}{\partial V}\left(\frac{\partial U}{\partial x} \times \frac{dx}{dV}\right) = \frac{\partial}{\partial x}\left(\frac{\partial U}{\partial x} \times \frac{dx}{dV}\right)\frac{dx}{dV}$$

$$= \frac{\partial^2 U}{\partial x^2}\left(\frac{dx}{dV}\right)^2 + \frac{\partial U}{\partial x}\left(\frac{d^2 x}{dV^2}\right) + \dots \quad (8.22)$$

At the equilibrium separation $(x = a)$ $\frac{\partial U}{\partial x} = 0$; consequently the second term vanishes so that

$$K = -V\frac{\partial^2 U}{\partial x^2}\left(\frac{dx}{dV}\right)^2. \quad (8.23)$$

We now relate V to the atomic spacing x. For a gram-molecule

$$V = 2N_0 x^3.$$

$$\therefore \quad \frac{dx}{dV} = \frac{1}{6N_0 x^2}. \quad (8.24)$$

To obtain K of equation (8.23) we differentiate equation (8.19) twice to obtain $\frac{\partial^2 U}{\partial x^2}$ and use equation (8.24) for $\left(\frac{dx}{dV}\right)^2$. We obtain

$$K = VN_0 1{\cdot}75 e^2 \left(\frac{-2}{x^3} + \frac{10a^8}{x^{11}}\right)\frac{1}{36N_0^2 x^4}. \quad (8.25)$$

At equilibrium when $x = a$ and $V = 2N_0 a^3$ we find

$$K = \frac{7{\cdot}00}{9} \times \frac{e^2}{a^4} = 0{\cdot}78\,\frac{e^2}{a^4}. \quad (8.26)$$

For rock salt, assuming complete charge separation on the ions, e is the electronic charge, $4{\cdot}8 \times 10^{-10}$ e.s.u. The equilibrium spacing $a = 2{\cdot}82 \times 10^{-8}$ cm. Substituting in equation (8.26) we obtain

$$K = 2{\cdot}8 \times 10^{11} \text{ dyne cm.}^{-2},$$

whereas the experimental value is about $2{\cdot}4 \times 10^{11}$ dyne cm.$^{-2}$ Taking into account the basic simplicity of the model this is surprisingly good agreement. It implies, of course, that the bonding is fully ionic. In most 'ionic' crystals only part of the bonding is ionic, the rest is covalent.

We may also express K in terms of the bond energy U_0 of the lattice in its equilibrium state. Substituting $x = a$ in equation (8.19) we have

$$U_0 = -N_0\frac{14{\cdot}0}{9}\left(\frac{e^2}{a}\right). \quad (8.27)$$

Hence from equation (8.26)

$$K = \frac{7 \cdot 0}{9} \frac{e^2}{a^4} = \frac{-U_0}{2N_0 a^3} = \frac{-U_0}{V},$$ (8.28)

where V is the volume of a gram-molecule. We shall see that a similar result is obtained for van der Waals solids.

It is interesting to note that, for a free NaCl molecule in the vapour state, the distance between the Na^+ and Cl^- ions is $2 \cdot 36$ Å. compared with $2 \cdot 82$ Å. in the crystal lattice. This indicates the important part played by the surrounding ions of opposite sign in opening up the ionic spacing. It is relatively easy to make a rough quantitative estimate of this. For a single ionic pair in the free molecule we may write

$$u = \frac{-e^2}{r} + \frac{A}{r^9}.$$ (8.29)

The equilibrium spacing r_0 occurs when $\frac{\partial u}{\partial r} = 0$. This gives the value of $A = \frac{r_0^8 e^2}{9}$. Hence

$$u = -e^2 \left(\frac{1}{r} - \frac{r_0^8}{9r^9} \right).$$ (8.30)

In the crystal the coulombic interaction of the ionic charges gives alternative positive and negative energies; as we saw above the final amount for each ion is $-1 \cdot 75 \, e^2/r$ where $1 \cdot 75$ is the Madelung constant. The repulsion term always involves a positive energy; however the potential falls off so rapidly with distance that we need only consider the 6 nearest neighbours at distance r for which the repulsion energy is $\frac{6A}{r^9}$ or $\frac{6r_0^8 e^2}{9r^9}$. The energy per ion is then

$$u = -e^2 \left(\frac{1 \cdot 75}{r} - \frac{6r_0^8}{9r^9} \right).$$ (8.31)

The minimum occurs when $\frac{\partial u}{\partial r} = 0$, i.e., when

$$\frac{1 \cdot 75}{r^2} = \frac{6r_0^8}{r^{10}},$$

$$r^8 = 3 \cdot 43 r_0^8 \quad \text{or} \quad r = 1 \cdot 17 r_0.$$ (8.32)

If r_0 is taken as $2 \cdot 36$ Å. this would give for r a value of $2 \cdot 76$ Å. which is close to the observed value. Even for a one-dimensional infinite chain molecule NaClNaClNaCl . . ., the ionic spacing is already increased from $2 \cdot 36$ Å. to $2 \cdot 69$ Å.

143   The elastic properties of solids

We consider a simple van der Waals solid. If it consists of single atoms or spherical or nearly spherical molecules it generally acquires a close-packed (face-centred cubic) structure. If the molecules are long chain hydrocarbons the packing tends to be very nearly hexagonal with the chains lying parallel to one another. We shall discuss only the face-centred cubic type of structure such as is found in solid argon, neon and krypton. For a pair of atoms at a distance x apart the van der Waals attractive forces produce a potential energy proportional to x^{-6}; the repulsion produces a potential energy which appears to be proportional to x^{-12}, rather than to x^{-9} as in the previous ionic model.

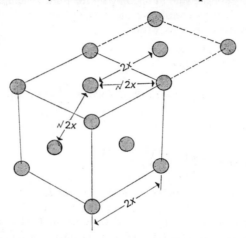

Figure 48.   The structure of a face-centred cubic structure of a solid inert gas. Each atom has 12 neighbours at distance $\sqrt{2}x$ apart, 6 neighbours at distance $2x$ apart, etc.

The potential energy between two atoms may be written

$$u = -\frac{A}{x^6} + \frac{B}{x^{12}}. \tag{8.33}$$

Each atom has 12 nearest neighbours distance $\sqrt{2}x$ apart and 6 neighbours $2x$ apart, etc. (see Figure 48). The contribution from the attraction forces is

$$-\left[\frac{12A}{(\sqrt{2}x)^6} + 6\frac{A}{(2x)^6} + \dots\right] = -\frac{1\cdot6A}{x^6} + \dots.$$

The convergence is so rapid that we may ignore further terms. The repulsion term may be grouped in a single constant C, so that the resultant potential energy may be written

$$u = -1\cdot6\frac{A}{x^6} + \frac{C}{x^{12}}. \tag{8.34}$$

144   Gases, liquids and solids

For N_0 atoms the total energy will be $\frac{1}{2}N_0$ times this. As before we may differentiate U to find the equilibrium condition when $x = a$.

$$U = -0.8N_0A\left(\frac{1}{x^6} - \frac{1}{2}\frac{a^6}{x^{12}}\right).$$ (8.35)

At the equilibrium separation

$$U_0 = -0.4\frac{N_0A}{a^6}.$$ (8.36)

The volume of the gram-atom for this structure is

$$V = 2N_0x^3,$$ (8.37)

so that as before

$$\frac{dx}{dV} = \frac{1}{6N_0x^2},$$ (8.38)

Repeating the previous procedure we find for the bulk modulus

$$K = \frac{1.6A}{a^9}.$$ (8.39)

The nearest-neighbour distance of atoms is $\sqrt{2}x$ and this is almost identical with the equilibrium separation r_0 for a single pair of atoms. We may thus rewrite equation (8.39) as

$$K = \frac{1.6}{a^3} \times \frac{8A}{r_0^6}.$$ (8.40)

The second factor in equation (8.40) has a very simple significance. By differentiating equation (8.33) we may determine B in terms of the equilibrium r_0 of one atom in relation to its neighbour. We shall then find the potential energy $\Delta\varepsilon$ of an isolated pair of atoms is simply $-\dfrac{A}{2r_0^6}$. Consequently from equation (8.40)

$$K = \frac{25.6\Delta\varepsilon}{a^3}.$$ (8.41)

For the inert gases and for the smaller molecules such as H_2 and N_2, $\Delta\varepsilon$ is of order $(5 \text{ to } 25) \times 10^{-15}$ erg, and a is about 3×10^{-8} cm. in the solid state. We obtain

$$K = (4 \text{ to } 20) \times 10^9 \text{ dyne cm.}^{-2}$$ (8.42)

This is about one hundred times smaller than for ionic solids.

145   The elastic properties of solids

We may also express K in terms of U_0. Combining equations (8.37) and (8.34) we have

$$K = -4 \frac{U_0}{Na^3} = -8 \frac{U_0}{V}. \tag{8.43}$$

This is similar to the result for ionic crystals except that the numerical factor is appreciably larger.

8·3·3 *Bond energy, or lattice energy, and heat of sublimation*

Before going on to discuss the bulk modulus of metals we make a few general comments on the bond (or lattice) energy and the heat of sublimation.

The total bond energy U_0 of a lattice (the lattice energy) is a fundamental quantity deriving from the forces between the constituent atoms and ions. It is not easily open to direct measurement. The nearest physical quantity is the heat of sublimation L_S, which in practice is a little greater than the sum of the latent heat of melting L_F and the latent heat of vaporization L_V. We may easily form an estimate of L_S for an ionic crystal. When sodium chloride is evaporated, the process does not involve pulling the positive and negative ions apart to infinity; this would be the true lattice energy. Vaporization generally produces individual NaCl molecules. L_S is thus less than U_0 by the value of the interaction energy between Na^+ and Cl^- in the individual NaCl molecule. From equation (8.31) the lattice energy per ion in the solid state is obtained by putting $r = 1 \cdot 17 r_0$ where r is the ionic spacing in the crystal, r_0 the spacing in the free NaCl molecule. The value is

$$u_{\text{lattice}} = -\frac{e^2}{r_0} (1 \cdot 33),$$

and for the gram-molecule

$$U_0 = -\frac{N_0 e^2}{r_0} (1 \cdot 33).$$

The energy per ion pair in the free molecule is obtained from equation (8.30) by putting $r = r_0$. We obtain for the free NaCl molecule

$$u_{\text{molecular}} = -\frac{e^2}{r_0} \left(\frac{8}{9} \right),$$

and for the gram-molecule

$$U_{\text{molecular}} = -\frac{Ne^2}{r_0} \left(\frac{8}{9} \right).$$

The molar heat of sublimation is therefore

$$L_S = U_0 - U_{\text{molecular}} = \frac{Ne^2}{r_0} (0 \cdot 44) \approx -\tfrac{1}{3} U_0. \tag{8.44}$$

The latent heat of sublimation, which we can measure with a fair degree of accuracy, is thus about one third of the lattice energy. Tables of values support this conclusion.

With a van der Waals solid sublimation merely separates all the molecules (or atoms) as individual molecules (or atoms). The heat of sublimation is thus a fairly accurate measure of the bond energy. We may easily form an estimate of this from the above analysis.

From equation (8.43) we have

$$U_0 = -\frac{kV}{8}$$

and from equation (8.41), this becomes

$$U_0 = \frac{3 \cdot 2 \Delta \varepsilon V}{a^3}. \tag{8.44a}$$

But from equation (8.37) we see that when the crystal is at its equilibrium separation $(x = a)$, $V = 2N_0 a^3$. Substituting in equation (8.44a) we have

$$U_0 = 6 \cdot 4 N_0 \Delta \varepsilon. \tag{8.44b}$$

This is the same as the value obtained in equation (7.2) except that the numerical factor is slightly different. This arises from the different assumptions made in the two models.

Metals present a very special problem. They consist of metal ions in a sea of free electrons. It is not possible to describe the forces between the ions and electrons in terms of a simple power law. Consequently the analysis we have given above is not possible. A reasonably good estimate of the bond energy U_0 can be made, but the treatment is very difficult and involves concepts which we shall not discuss in this book. Even if we knew the bond energy it would not correspond to the latent heat of sublimation. The bond energy would correspond to the work done in separating all the ions and all the free electrons to infinity; whereas in the vapour state the metal exists essentially as free atoms. Thus $U_0 = L_S +$ ionization energy of the metal atoms. A similar type of difficulty exists with covalent solids.

8·3·4   *Bulk modulus of metals*

For the reasons given above it is not possible to derive the elastic properties of a metal using the simple type of analysis that we have applied to ionic and van der Waals solids. However we may form a reasonable estimate of the bulk modulus by noting that, in general,

$$K = -c\frac{U_0}{V}, \tag{8.45}$$

where U_0 is the total bond energy per gram-atom (or molecule), V the atomic (or molecular) volume and c a constant depending on the law of forces. For an ionic crystal $c = 1$; for a van der Waals solid it is 8. For metals we can only guess at a value of c. We also have to recognize that we have no simple way of calculating U_0, but we can use the sublimation energy L_s (which can be determined experimentally) as some measure of U_0. As a matter of interest we shall try the relation

$$K = 3 \frac{L_s}{V} \tag{8.46}$$

where 3 is purely an empirical factor. This represents an attempt to allow both for the unknown law of force between metal ions in the lattice and for the discrepancy between the bond energy and the latent heat of sublimation. For L_s we have simply added the latent heats of melting and of vaporization (taken from C. J. Smithells (1962), *Metals Reference Book*, 3rd edn, Butterworth) and have ignored any specific heat terms. The results are given in Table 21.

Table 21 *Latent heat of sublimation L_s and bulk modulus K of some metals*

Metal	L_s per g-atom		Atomic vol. V	$K \times 10^{11}$ dyne cm.$^{-2}$	
	Kcal.	erg $\times 10^{10}$	cm.$^{-3}$	calc.	observed
K	22	92	45·5	0·6	0·4
Mg	36	152	14·0	3·3	3·6
Pb	47	198	17·7	3·4	4·6
Al	77	322	10	9·7	7·5
Cu	82	345	7	15	13·8
Fe	95	400	7·1	17	17·8
Mo	159	670	9·4	21	26·1
W	203	850	9·5	27	31·1

The agreement is surprisingly good in view of the over-simplifying assumptions made.

8·4 Elastic properties of rubber

The elastic properties of rubber are very different from those of most other solids such as ionic solids, covalent solids and metals. In tension or shear the modulus is 1000 to 10,000 times smaller. On the other hand Poisson's ratio is almost $\frac{1}{2}$, so the bulk modulus is relatively high; it approaches that of a liquid. The two most striking elastic features are (*a*) rubber has enormous extensibility – up to 400 per cent compared with perhaps 1 per cent for ordinary solids, (b) the elastic modulus increases with increasing temperature. This is the only solid known which shows such an effect. We recognize that these two properties are very similar to the behaviour of a gas in compression.

Since rubber is known to consist of long chain organic molecules we realize at once that its enormous extensibility and low modulus cannot arise from the stretching or bending of the C—C bonds, since these are extremely strong and difficult to deform. The chains are, in fact, coiled up in higgledy-piggledy fashion and the result of applying a tension is to pull the chains into partial alignment. Thermal energy attempts to increase the disorder in the chain and therefore resists the aligning action of the tensile force. Consequently the modulus in tension increases with temperature.

If the molecular chains were completely separate the material would be a very viscous liquid and indeed natural latex is such a liquid. In order to become a rubbery solid the molecular chains must be cross-linked at a relatively small number of points. We now consider briefly the properties of a typical segment lying between two cross-links. Suppose the C—C bond has a length λ and there are N such bonds in a segment; because *rotation* about the C—C bond, in contrast to stretching or bending, is extremely easy, as a first approximation we may ignore any stiffness which arises from this factor. The segment then consists of N freely pivoted bonds. We may then ask: what is the most probable length of a segment of N units each of length λ?

Assuming that each link is completely free to take up any position relative to its neighbour* we recognize this as the problem of a 'random walk' in three dimensions or a 'random flight'.

We indicate briefly here how this may be calculated. Consider first a random walk in one dimension. Suppose a man takes N steps, each of length λ, along a line; each step may be forwards or backwards. What is the probability that he will have travelled a distance x from his starting point? If out of N steps A are forwards and B are backwards

$$A+B = N$$

and

$$A\lambda - B\lambda = x. \tag{8.47}$$

Hence

$$A = \tfrac{1}{2}\left(N+\frac{x}{\lambda}\right)$$

$$B = \tfrac{1}{2}\left(N-\frac{x}{\lambda}\right). \tag{8.48}$$

Any way of taking N steps such that A are forwards and B backwards will give

---

* This ignores the requirement of fixed bond-angles between the carbon atoms. If this is taken into account, and rotations about the bonds are permitted, the value of β in equation (8.54) is merely changed by a small numerical factor. A more serious conceptual difficulty is that in the random walk any given step may be covered many times, whereas in a real molecule the location of a molecular link can only be occupied once.

the resultant distance x. The number of ways W in which this can be achieved is

$$W = \frac{N!}{A!\,B!}.\qquad(8.49)$$

For simplicity assume N to be even so that A and B [from equation (8.48)] are integers. We use the Stirling approximation $N! = \left(\frac{N}{e}\right)^N$ and take $x \ll N\lambda$ (in the rubber molecule this is equivalent to assuming that the distance between the two ends of the segment is very much less than the fully extended length of the segment). Then

$$\ln W = N\ln 2 - \frac{x^2}{2N\lambda^2}$$

or

$$W = Be^{-\beta_1^2 x^2}.\qquad(8.50)$$

where $\beta_1^2 = \frac{1}{2N\lambda^2}$. The probability P($x$) of the distance x occurring is directly proportional to W so that

$$P(x) = Ce^{-\beta_1^2 x^2}.\qquad(8.51)$$

The constant C may be determined by observing that

$$\sum_{x=-N\lambda}^{x=N\lambda} P(x) = 1 \simeq \int_{-\infty}^{\infty} P(x)dx.\qquad(8.52)$$

This gives $C = \frac{\beta_1}{\pi^{\frac{1}{2}}}$.

We now turn to the three-dimensional random walk. If the coordinates of the starting point are 0, 0, 0 and those of the final point are x, y, z on orthogonal axes, the probability of finding the final position is

$$P(x, y, z) = \frac{\beta^3}{\pi^{\frac{3}{2}}}e^{-\beta^2(x^2+y^2+z^2)}.\qquad(8.53)$$

Normalizing for three dimensions it turns out that the β^2 in equation (8.53) is 3 times β_1 for the one-dimensional random walk, i.e.

$$\beta^2 = \frac{3}{2}\frac{1}{N\lambda^2}.\qquad(8.54)$$

We are only interested in the exponential factor. We may therefore write for the most probable length of a segment

$$P(r) = Ae^{-\beta^2 r^2}.\qquad(8.55)$$

150 Gases, liquids and solids

If one end of the segment is fixed, the probability of finding the other end in an element of volume $d\tau$ is

$$Pd\tau = Ae^{-\beta^2 r^2} d\tau. \tag{8.56}$$

We see that the probability density P is a maximum when $r = 0$, that is the segment makes its greatest effort to have zero distance between its ends. This is to be expected since if all steps in the random walk are equally probable a large number will contain as many forward as backward ones. This does not mean that the most probable distance between chain ends is zero. If we are interested only in the length r and not the direction then (as in our treatment of molecular velocities in a gas) we may consider one end of the segment fixed and determine the number of the other ends lying in a shell of volume $4\pi r^2 dr$. Then the number Π with lengths between r and dr is

$$\Pi dr = A4\pi r^2 e^{-\beta^2 r^2} dr.$$

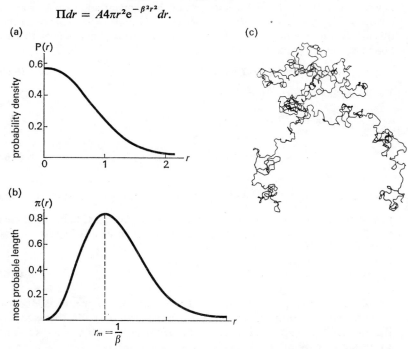

Figure 49. (a) Probability density P as function of distance between the ends of a long chain assuming a three-dimensional 'random walk', (b) most probable distance r between the ends of a chain with N segments each of length λ occurs for $r_m = \dfrac{1}{\beta} = \lambda\sqrt{(\tfrac{2}{3}N)}$, (c) random arrangement of a chain containing 1000 segments, each segment set at the appropriate bond angle, but being given a random choice of 6 equally-spaced angular positions: this is a reasonably close approximation to complete randomness (from L. R. G. Treloar (1958), *Physics of Rubber Elasticity*, 2nd edn, Oxford University Press).

151   The elastic properties of solids

The result is shown in figure 49(*b*); the maximum occurs for a value of *r* given by

$$r_m = \frac{1}{\beta} = \lambda\sqrt{(\tfrac{2}{3}N)}. \qquad (8.57)$$

A simple analogy is to consider the concentration of bees around a drop of honey. The greatest density occurs near the honey but the greatest number of bees may well be found between 5 and 6 cm. from the drop, rather than in the first cm.

Reverting to equation (8.55) the probability density is *P* and this is related to the (configurational) entropy of the segment by the relation

$$S = k \log P = \text{constant} - k\beta^2 r^2. \qquad (8.58)$$

If we apply a tensile force *f* to one end of the segment and extend it in the direction of *r* by an amount *dr*, the external work done is *f dr*. Assuming that this involves no volume change we may equate it to the change in the Helmholtz free energy [equation (3.23)].

$$f\,dr = dA = d(U - TS). \qquad (8.59)$$

For an isothermal extension

$$f = \frac{\partial U}{\partial r} - T\frac{\partial S}{\partial r}$$

$$= \frac{\partial U}{\partial r} + 2kT\beta^2 r. \qquad (8.60)$$

The term $\dfrac{\partial U}{\partial r}$ refers to the change in *internal* energy with extension, that is, with uncoiling of the segment by rotation about the C—C bond. As there is very little energy change (in our model no change at all) in this we may neglect this term and write

$$f = 2kT\beta^2 r = \frac{2RT\beta^2 r}{N_0}. \qquad (8.61)$$

We see at once that (*a*) *f* is proportional to *r* so that the segment has Hookean properties, (*b*) *f* is proportional to *T*, (*c*) *f* is proportional to β^2, that is to $1/N$: the smaller the number of bonds in the segment, the greater its resistance to elastic extension.

So far we have only discussed the force that the segment exerts in resisting extension; whatever its length it is always attempting to contract. In bulk rubber which consists of a network of segments and chains it turns out that even if the rubber is subjected to uniaxial compression the overall effect, because of the large Poisson's ratio, is to extend the chains. The specimen therefore resists both extension and compression in a manner resembling the behaviour of a single segment discussed above.

We see that the elasticity of rubber arises from the entropy or randomness of the chain segments. It is for this reason that it resembles the elasticity of an ideal gas in compression where the pressure may be considered as arising from the reduction in entropy associated with a reduction in volume. Indeed one notices the presence of the 'gas constant' R as a fundamental term in the expression for rubber elasticity in equation **(8.61)**. In view of this one might well designate R as the 'rubber constant' in 'ideal rubber' elasticity.

Chapter 9
The strength properties of solids

9·1 Deformation properties

9·1·1 *Ductile properties*

The strength properties of solids are most simply illustrated by considering the behaviour of a homogeneous specimen of uniform cross-section subjected

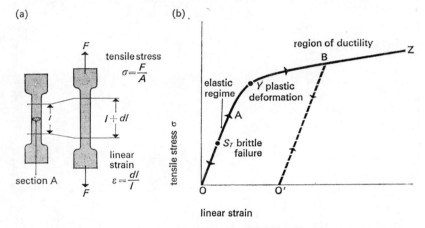

(a)

(b)

F

tensile stress

$$\sigma = \frac{F}{A}$$

$l + dl$

linear strain

section A

$$\varepsilon = \frac{dl}{l}$$

F

region of ductility

B

Z

elastic regime

Y plastic deformation

A

S_T brittle failure

tensile stress σ

O

O′

linear strain

Figure 50. (*a*) Tensile specimen and (*b*) typical stress–strain curve, showing elastic deformation along OA, plastic yielding at Y and work-hardening along YZ A brittle solid fails at a tensile stress *S*.

to uniaxial tension [figure 50(*a*)]. If we plot the true stress σ against the linear strain ε (i.e., the fractional increase in length) we may obtain a curve as illustrated in figure 50(*b*). The portion OA represents elastic deformation. The strain is proportional to the stress and the deformation is reversible. If the material is ductile, elastic deformation will proceed until at some critical stress *Y* the onset of permanent or plastic deformation occurs. If we continue along the plastic curve there is generally an increase in yield stress with deformation. This is known as work-hardening. If at some point B we reduce the stress the material recovers elastically along BO′, where BO′ is very nearly parallel to OA. The displacement OO′ is the permanent plastic extension produced in the specimen. On reapplying the stress the deformation follows the curve O'BZ.

Experiments show that under simple uniaxial compression a cylinder will first deform elastically and then yield plastically at the same compressive stress Y.

Some insight into the yield criterion is provided by considering the effect of hydrostatic pressure, that is a stress situation in which the comprehensive stress on the material is the same in all directions. If we subject the specimen to a hydrostatic pressure P we find that plastic flow does not occur even if P exceeds Y. We must still apply a uniaxial stress (either in tension or compression) and its magnitude if plastic flow is to occur is still Y. Now analysis shows that the only part of a stress field which is unaffected by hydrostatic pressure is a shear stress. We conclude that plastic flow is associated with a critical shear stress. This is fully supported by microscopic studies which show that plastic deformation is always accompanied by slip of atomic planes over one another.

9·1·2 Shear stress

Consider a rectangular bar of uniform cross-sectional area A [figure 51(a)]. Suppose we apply a tensile force F, so that the stress is $\sigma = F/A$. Consider a thin slice of material making an angle θ to the direction of F. On one side of

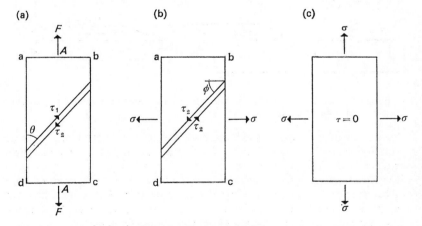

Figure 51. Shear stresses produced by tensile stresses: (a) vertical stress σ, (b) horizontal stress σ, (c) two-dimensional hydrostatic stress σ; this produces a resultant zero shear stress.

the slice there is a force $F \cos \theta$, on the other side a force of equal magnitude in the opposite direction. These forces constitute a shear. The surface area

155   The strength properties of solids

of the slice is $A/\sin\theta$ so that the shear stress is given by

$$\tau_1 = \frac{F\cos\theta}{\dfrac{A}{\sin\theta}} = \frac{F}{A}\cos\theta\sin\theta$$

$$= \sigma\cos\theta\sin\theta. \tag{9.1}$$

The maximum occurs for $\theta = 45°$ and has a value $\tau = \dfrac{\sigma}{2}$.

If we apply a tensile stress σ to the faces bc, da, the shear stress on the slice is

$$\tau_2 = \sigma\cos\phi\sin\phi$$

$$= \sigma\sin\theta\cos\theta. \tag{9.2}$$

This stress is equal to τ_1 and in the opposite direction. If the two systems of tensile stresses σ are superposed (to constitute a two-dimensional hydrostatic tension), the two shear-stress systems completely annul one another. This is the basis for the statement that hydrostatic stresses do not in any way change the existing shear stresses in a system.

Finally we note that the conventional method of representing a shear stress by two equal and opposite parallel stresses τ is misleading [figure 52(a)].

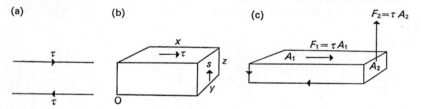

Figure 52. Sketch showing that a shear stress must always involve a pair of equal orthogonal shear stresses.

Such a stress system would produce a couple which would produce continuous rotation of the stressed element. There must be an opposing set of shear stresses s if static equilibrium is to be achieved. Consider a rectangular element length x, width y, depth z. On the upper and lower faces we apply shear stresses τ; on the side faces shear stresses s. For equilibrium take moments about O. Shear stress τ implies a shear force $\tau \times xy$; its couple is $\tau \times xy \times z$. Similarly shear stress s implies a couple $s \times yz \times x$.

Then $\tau xy \times z = s \times yz \times x$,

$$\tau = s. \tag{9.3}$$

We see that a shear stress always involves a pair of equal orthogonal stresses. In practice, of course, one is often unaware of this. For example in shearing a long thin slice as in figure 52(c), the horizontal shear force $F_1 = \tau A_1$ may be very large and the vertical shearing force $F_2 = \tau A_2$ may be vanishingly small if A_2 is very small, although the value of the shear stress τ must be the same in both cases.

A tensile stress Y produces a maximum shear stress $Y/2$ at 45° to the direction of Y. (The same applies to a uniaxial compressive stress Y.) For an isotropic material shear will therefore occur in slip directions at 45° to the direction of the applied stress. If the material is not isotropic shear may occur more easily in some directions than others; if the shear stress is exceeded in this direction slip will occur in these more favourably orientated directions. This is often observed with single crystals, and is shown schematically for a simple shear stress in figure 53.

(a)   (b)

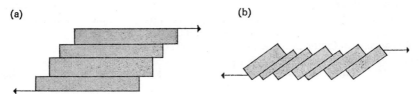

Figure 53.   The effect of a shear stress on the yielding of a single crystal (a) when the shear plane is parallel to the shear, (b) when it is inclined.

We may now explain the behaviour shown in figure 50(b). When the stress is first applied the atoms are displaced from their equilibrium positions; the resistance to deformation is determined by the interatomic forces. When the tensile stress reaches a critical value the shear stress is sufficient to produce slip along an appropriate plane and plastic yielding occurs. However the whole stress must still be supported by the displaced atoms. Consequently when the stress is removed there is elastic recovery and the modulus, which arises from the interatomic forces, is essentially the same as originally.

9·1·3   Indentation hardness of ductile solids

Engineers and metallurgists often wish to determine the strength properties of their materials without going to the trouble of preparing tensile specimens and carrying out full-scale tensile tests. A very convenient way of doing this is to measure the indentation hardness. A very hard indenter (a hard steel sphere in the Brinell test, a diamond pyramid in the Vickers test) is pressed under a load W into the surface of the material to form a plastic indentation. When the indenter is removed the diameter of the indentation is measured and its area A determined. The mean pressure over the indentation is then

$$p = \frac{W}{A}. \tag{9.4}$$

In the industrial test procedure A is the surface area of the indentation; from the point of view of making p physically meaningful, the projected area is more appropriate. The difference however is generally small. A study of the stress situation around the indenter shows that almost two-thirds of p is in the form of a hydrostatic component, and therefore plays no part in producing plastic flow. Consequently only one-third of p is active in producing the indentation. Thus as a first approximation

$$p = 3Y, \tag{9.5}$$

where Y is the uniaxial yield stress of the material. This relation is well substantiated in practice.

Further, if the material does not appreciably work-harden after yielding in tension its yield stress Y will be very nearly the maximum stress the material can support before it pulls apart, i.e., its ultimate tensile strength (u.t.s.). The latter is therefore about one-third the indentation hardness. Since p is normally measured in kg. mm.$^{-2}$ and the u.t.s. in ton in.$^{-2}$ the conversion factor must be divided by 1·58. Consequently

$$\text{u.t.s. (ton in.}^{-2}) = \frac{0 \cdot 33}{1 \cdot 58} p = 0 \cdot 21 p \text{ (kg. mm.}^{-2}). \tag{9.6}$$

This conversion ratio is widely used for polycrystalline homogeneous materials and is reliable to within a few per cent.

Typical values for the indentation hardness in kg. mm.$^{-2}$ are 0·25 for krypton at $-220°C$., 0·8 for ice at $-10°C$., 1 for indium, 4 for lead, 40 for polycrystalline copper, 120 for mild steel, 900 for ball-bearing steel, 2000 for sapphire and over 10,000 for diamond at room temperature.

9·1·4   *Calculation of critical shear stress for single crystals*

We now consider the shear stress at which two neighbouring planes in a single crystal can be caused to slide over one another. Consider the arrangement in figure 54 for a typical face-centred cubic crystal and apply a shear stress τ to the planes X and Y. If θ is the strain produced and G is the rigidity or shear modulus

$$\tau = G\theta.$$

Atom B in its initial site in the lattice is in a position of minimum potential energy. As it is displaced the P.E. increases until a position of unstable equilibrium is reached at B′ – the P.E. curve at this point is horizontal but is a maximum. The P.E. curve thus has the form shown in figure 54(b). The force curve is found by differentiating the P.E. curve: it is drawn in figure 54(c) and shows that the maximum force occurs when B has reached some intermediate position B″ between B and B′. The angle of strain at this point is about $\frac{1}{4}$. If G remained constant we could say that B is sheared over to B′ for a shear stress of magnitude

$$\tau_m \simeq \frac{G}{4}. \tag{9.7}$$

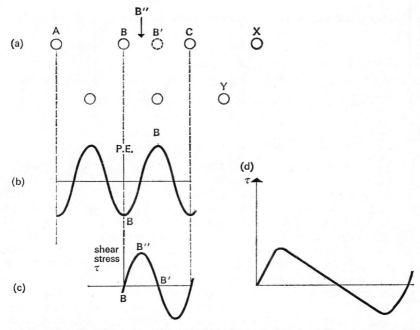

Figure 54. Shearing of a perfect crystal. (*a*) Neighbouring planes of atoms at X and Y, (*b*) the potential energy curve as the top plane is slid over the bottom, (*c*) the equivalent force–displacement diagram, (*d*) a more realistic form of the force–displacement curve.

The lattice could not resist a greater stress than this so that plane X would glide over plane Y for this value of τ.

This is an overestimate since it assumes that G is a constant and has the same value as for small strains. This is rather unrealistic since for larger distortions we expect the modulus to decrease. We may form a better estimate of τ_m by assuming that the shear stress is a sine function of the displacement. It must clearly be zero for displacements $x = 0$, $x = a/2$ and $x = a$, where a is the atomic spacing. We may thus use a function of the form

$$\tau = A_0 \sin \left(\frac{2\pi}{a} x\right), \tag{9.8}$$

where A_0 is a suitable constant. We may determine A_0 by specifying that the initial slope near $x = 0$, where the strains are small, corresponds to the shear modulus G of the material. For small x

$$\tau = A_0 \sin \frac{2\pi}{a} x = A_0 \frac{2\pi}{a} x = G\theta, \tag{9.9}$$

when θ is the shear angle $\dfrac{x}{a}$. Consequently

$$A_0 \frac{2\pi}{a} x = G \frac{x}{a} \quad \text{or} \quad A_0 = \frac{G}{2\pi}. \tag{9.10}$$

The shear stress reaches its maximum when the displacement is midway between B and B'; here $\sin\left(\dfrac{2\pi}{a} x\right)$ is unity so that

$$\tau_m = A_0 = \frac{G}{2\pi}. \tag{9.11}$$

A stress of this value will carry atom B over to position B'' and then for a slight increase in τ it will flick over to B' and then to C. Thus slip along the atomic plane will take place for a shear stress of about $G/6$. A more realistic study of the atomic force field suggests a curve like that shown in figure 54(d). This reduces the critical shear stress τ_m even further but no amount of adjustment can reduce it to a value less than about $G/30$.

We conclude that the maximum shear stress τ_m the lattice can withstand is of the order of $\frac{1}{10}$ to $\frac{1}{30} G$. Before discussing the relation of (τ_m) theoretical to (τ_m) experimental we may ask a very simple but pertinent question. Once we have dragged atom B to the top of the potential hill at B' [figure 54(b)] why does it not glide over the rest of the potential hills until planes X and Y have completely slid off one another? Is not the position analogous to a 'frictionless' helter skelter, where once the carriage has reached its highest point it can proceed indefinitely over all hills which do not exceed that height? The simple answer is that such a model would be appropriate if the atoms in row Y were absolutely rigid in space. Because they are themselves part of an elastic system the behaviour is much more like the interaction of two (frictionless) combs the teeth of one being dragged through those of the other [figure 55(a)]. Up to a certain tangential stress both sets of teeth are distorted reversibly [figure 55(b)]; beyond this point the teeth from the upper comb

(a)   (b)   (c)

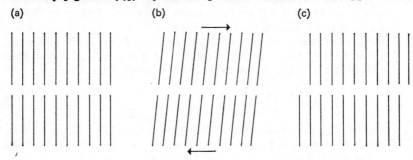

Figure 55. The displacement of the teeth of two sets of engaging combs (a) initially, (b) after a small stress has been applied, (c) after the stress has been sufficient to cause slip: the elastic strain energy in the teeth is lost by vibration.

escape from interaction with the lower comb and flick over into their new equilibrium position; whilst the teeth of the lower comb flick back into their original position [figure 55(c)]. The whole of the distortional energy has disappeared as vibration of the teeth. This has a close analogue in the slip of crystals. Practically the whole of the work of plastic deformation is dissipated as vibrational energy in the lattice, that is as heat. Not more than a few per cent of the energy is retained as strain energy in the lattice. This has an interesting corollary in an unexpected area. The friction of metals is largely due to adhesion, shearing and deformation within and around the regions of real contact; it is, in fact, a process which is generally dominated by plastic flow around the contact zones. The plastic work virtually all appears as heat. This is the reason for Joule's observation that in frictional heating there is quantitative agreement between the work done against friction and the heat liberated. If large amounts of energy could be stored in the lattice this equivalence would not be observed.

9·2 Dislocations

9·2·1 *Observed shear stress, need for dislocations*

Experiments on pure single crystals show that minute shear stresses are sufficient to produce permanent, i.e. plastic, deformation. In general (τ_m) observed is of the order of one thousand times smaller than (τ_m) theoretical. The theoretical value cannot be wrong to this extent. This discrepancy has led to a search for sources of weakness in the lattice, such as flaws or imperfections: the simplest type is that known as the edge dislocation shown in figure 56(a). It involves, essentially, the insertion of an extra half-

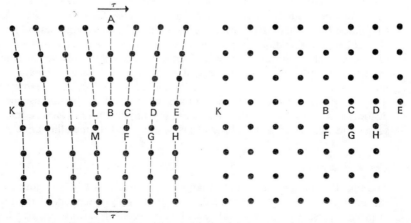

Figure 56. (a) An edge dislocation, (b) the displacements which occur when a shear stress is applied and the dislocation at B moves out to E.

161 The strength properties of solids

plane of atoms AB into the lattice. If a shear stress is applied as shown, atom B flicks over and joins with atom F for a very small value of τ. Then C joins up with G and D with H so that for a very small stress we have produced slip by one atomic spacing DE [figure 56(b)]. In effect what we are saying here is that the theoretical value of τ_m assumes that slip occurs simultaneously across the whole plane KE; in the dislocation model slip is able to occur one row of atoms at a time and this leads to an enormous reduction in shear stress.

We can already see that there are two basic problems associated with the above model. First, how are the dislocations originally produced in the crystal? Secondly when the edge dislocation has slipped out of the crystal as in figure 56(b), the crystal has become perfect so that it has its ideal theoretical strength. As this is not generally observed there must be some process for the generation of further dislocations whilst slip is occurring. Both these questions are part of a very large subject which we shall not pursue further in detail. We mention however a few further points of interest.

9·2·2   *Direct experimental evidence for dislocations*

The first and most direct evidence for the existence of dislocations is due to Dr. J. W. Menter, who studied, with an electron microscope, very thin specimens of platinum phthalocyanine. This is a metallo-organic compound which forms well-defined crystals in which the platinum atoms lie on crystal planes 12 Å. apart. In the electron microscope the organic part of the crystal scarcely absorbs electrons whilst the platinum atoms strongly absorb. In transmission, therefore, a microscope with a resolution of say 10 Å. is able to resolve the planes of the platinum atoms. Typical election micrographs [see J. W. Menter (1956), *Proceedings of the Royal Society*, A236, p. 119, and F. P. Bowden and D. Tabor (1964), *Friction and Lubrication of Solids*, Clarendon Press, Plate III] reveal the presence of a defect strongly resembling a classical edge dislocation. In pure metals the atomic spacing is of the order of 3 Å. and electron microscopes are not generally able to resolve distances smaller than about 5 Å. Consequently it is not possible to 'see' dislocations in metals in the direct way that they are revealed in phthalocyanine. However, it is now possible to identify certain features revealed by the microscope as dislocations. This has led to a very extensive application of electron microscopy and diffraction to the study of dislocations in thin metallic films.

9·2·3   *Upper yield point*

The edge dislocation is a region of large elastic stress. At LBC in figure 56(a) the material is heavily compressed, at MF strongly extended. If impurity atoms are present in the lattice they will tend to concentrate at regions where they can relieve the stress. For example a *small* impurity atom will tend to take the place of the original atom B. This will lower the intensity of the stress field around the dislocation. It is precisely the presence of the stress field that makes it easy for a dislocation to move. The presence of the impurity at B thus tends to 'pin' the dislocation and a larger stress than normal will be required to make it move. Once the dislocation has moved across to position C or D it is in its normal environment and is now much easier to move. This explains the behaviour for example of certain mild steels where the dislocations are pinned by interstitial carbon or nitrogen. When such materials are tested in tension or compression or shear they show an 'upper yield point' (see

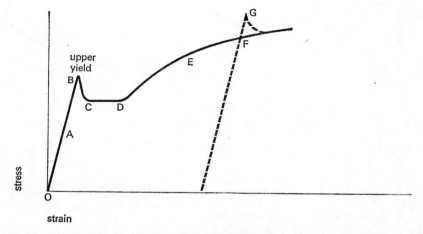

Figure 57. Schematic diagram showing upper yield stress at B and a second upper yield at G which occurs after 'ageing'.

figure 57, point B), then a region of fairly constant lower yield stress (CD) followed by a conventional work-hardening curve. If at some point F the stress is removed and the metal is 'aged' the carbon and nitrogen can diffuse to the dislocations and again pin them. The further deformation of the specimen may then show a second 'upper yield point' (G).

9·2·4 Work-hardening

Dislocations interact with one another. Continued deformation produces interlocking of dislocations and makes it more difficult for them to move. Although the details are still the subject of discussion this is the basic reason for work-hardening as a result of deformation.

9·2·5 Screw dislocations

Apart from edge dislocations there are 'screw' dislocations as shown in figure 58(*a*). If we follow a sequence of atoms, say, in the top plane starting

163   The strength properties of solids

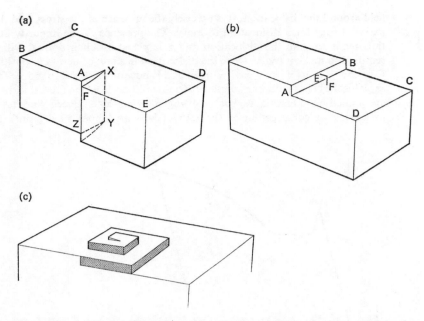

Figure 58. (*a*) Screw dislocation, (*b*) crystal growth on an edge providing two points of attachment at E and F, (*c*) crystal growth on the edge of a screw dislocation which provides a self-perpetuating growth step.

from A and moving to B C D E F we end up at the point F which is an integral number of atomic spaces below A. This is essentially a screw movement. If the dislocation line XY moves towards CD the slipped region AXYZ increases in area.

9·2·6 *Crystal growth*

The presence of sites for nucleation is very important in crystal growth. On a perfectly smooth atomic plane an atom can find only one point of attachment. On a step, each atom can find two points of attachment [figure 58(*b*)]. Growth will therefore occur very much more easily until the whole of the step ABCD is filled, when the surface is then plane. A screw dislocation, however, provides a self perpetuating growth step; as further atoms become attached to the step they build up a spiral 'staircase' and so provide further steps for further condensation. This mechanism accounts for the growth spiral that is often observed on the face of crystals. [see figure 58(*c*)]

9·3 Brittle solids

9·3·1 *Brittle properties*

We now consider the behaviour of a solid which normally shows no ductility. If it is extended it stretches elastically and then snaps at some critical tensile stress S. If the material is homogeneous it will crack along a plane normal to the direction of the stress. If the specimen is subjected to a hydrostatic pressure P the tensile stress necessary to cause brittle failure is found to be $P+S$ (figure 59). This implies that brittle failure occurs when the resulting

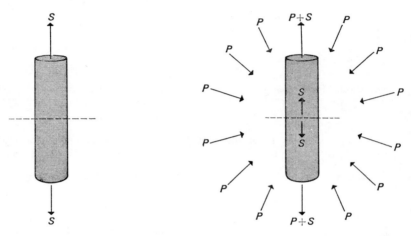

Figure 59. Brittle failure: (*a*) in pure tension the specimen fails for a tensile stress S, (*b*) when a hydrostatic pressure P is applied brittle failure occurs for a tensile stress of $P+S$.

tensile stress across some appropriate plane exceeds a critical value, in this case S. In its idealized form this is the stress necessary to pull one plane of atoms completely away from a neighbouring plane.

We may estimate the theoretical brittle strength in a number of ways. If the bar has unit cross-section and γ is the surface energy, the act of snapping the specimen is to create two new unit areas of surface: the work done is 2γ. If intermolecular forces are appreciable only over the distance of an atomic spacing, say 5 Å., and the stress over this distance is assumed constant, we may write

$$\text{work done} = S \times 5 \times 10^{-8} = 2\gamma. \tag{9.12}$$

For paraffin wax, where γ is of the order of 100 erg cm.$^{-2}$ this gives a value of S of about 40×10^8 dyne cm.$^{-2}$ or 4000 kgf. cm.^{2} For rock salt for which γ is about 500 erg cm.$^{-2}$ the theoretical value is of order 20,000 kgf. cm.^{2} This is enormously larger than the observed strength.

165   The strength properties of solids

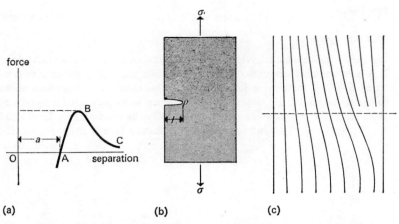

Figure 60. (a) Force–displacement curve for calculating the theoretical brittle strength. (b) role of a crack, length l, tip radius ρ, in facilitating brittle failure, (c) the formation of a crack by the 'pile-up' of dislocations.

We may also estimate the strength from the force–separation curve [figure 60(a)]. When the force reaches the point B it is great enough to pull the atoms apart. This occurs when $\dfrac{\partial F}{\partial x_i} = 0$. If we express the energy U in terms of x we have $\dfrac{\partial U}{\partial x} = -F$. Hence failure occurs when

$$\frac{\partial^2 U}{\partial x^2} = 0. \tag{9.13}$$

We have not derived an expression for the energy of a crystal in terms of the atomic separation in one dimension. We may however estimate the hydrostatic tension necessary to pull a crystal apart. For an ionic crystal we have (from the previous chapter)

$$U = -\text{const} \left[\frac{1}{x} - \frac{a^8}{9x^9} \right]. \tag{9.14}$$

Ignoring the constant and differentiating,

$$\frac{\partial^2 U}{\partial x^2} = -\frac{2}{x^3} + \frac{10a^8}{x^{11}} = 0$$

$$x = 5^{\frac{1}{8}} \times a = 1 \cdot 23a. \tag{9.15}$$

This shows that the crystal can be expanded by a hydrostatic tension until when the ionic separation has increased by 23 per cent, i.e., the volume has increased by 85 per cent, it will disintegrate. If the bulk modulus K remained constant for these large strains it would imply a pressure equal to 0·85 K, but because of the large curvature of the force–separation curve in this region the stresses are appreciably smaller. Simple calculation shows that it should be reduced by a factor of about 3·5 so that the corresponding pressure is about 0·24 K. Using the theoretical value of K from the previous chapter, this is 7×10^{10} dyne cm.$^{-2}$ or 70,000 kgf. cm.^{2} This is of the same order as that deduced from the surface energy. In practice a hydrostatic tensile experiment would be difficult if not impossible to carry out; most estimates based on direct tension give a theoretical value for the brittle strength of the order of

$$S \simeq \frac{E}{10},$$ (9.16)

where E is Young's modulus.

9·3·2   *Observed brittle strength: need for cracks*

The observed brittle strength of solids is generally very variable but is always 10 to 100 times smaller than the theoretical value. The source of this weakness is the presence of flaws or cracks in the solid, especially at the surface. These act as 'stress-raisers' or 'stress-multipliers'. Consider a rectangular strip of uniform thickness subjected to a tensile stress σ. If there is a crack in one edge, as shown in figure 60(*b*), and we can specify it by its length l and its tip radius ρ the tensile stress at the tip of the crack is multiplied by a factor of the order

$$\sqrt{\frac{l}{\rho}}.$$ (9.17)

Hence if the applied stress is σ and the theoretical breaking stress is S, we may write

$$\sigma \simeq S \sqrt{\frac{\rho}{l}}.$$ (9.18)

If the crack radius is, say, 10 Å. a crack length of 10,000 Å. will increase the tensile stress at the tip of the crack by 30-fold. Thus an applied stress only one-thirtieth of the theoretical strength will be able to start the crack growing. Once this occurs the whole section will fail.

There are three general observations which support this explanation. First brittle strengths are usually very variable; this is because of the large variety of sizes and shapes of surface cracks. Secondly, in recent years it has proved possible to prepare fine silica rods with very perfect surfaces which can withstand a tensile stress of $E/20$ before they fail. Thirdly, some fine crystalline fibres have been prepared which have tensile strengths approaching their theoretical value, presumably because their surfaces are free of surface flaws or growth steps.

Although this explanation is generally satisfactory there is an additional failure criterion which must be mentioned. Consider a flaw which is in the form of a sharp crack between two neighbouring atomic planes; the crack-tip radius tends to zero. Is then the stress-concentration factor infinite, and does the material now possess zero strength? The answer was given by A. A. Griffith in 1921. He showed that for a crack to grow it is not sufficient for the stresses at the crack-tip to exceed the theoretical strength; in addition sufficient elastic energy must be released from the system to provide the extra surface energy that a growing crack demands. By expressing the surface energy in terms of the elastic constants, the force–separation curve, and the atomic spacing it is possible to estimate what this involves. It turns out [see A. H. Cottrell (1964), *The Mechanical Properties of Matter*, Wiley, London, and E. Orowan (1949), 'Fracture and strength of solids', *Reports on Progress in Physics*, vol. 12, 1949, pp. 185–232], that an infinitely sharp crack cannot grow with the application of a vanishingly small tensile stress. In order to satisfy the surface energy criterion, the smallest stress σ_g capable of producing crack propagation, if the lattice spacing is a, is of order

$$\sigma_g \simeq S \sqrt{\left(\frac{3a}{l}\right)}, \tag{9.19}$$

that is to say, because of the surface energy criterion an infinitely sharp crack produces a reduction in strength equivalent to a tip radius of about three atomic spacings. For a crack of length l, the Griffith stress σ_g is the smallest stress that will start the crack growing. This is the stress for a crack of tip radius $3a$ or less. For a more blunt crack ($\rho > 3a$) a larger stress is needed to start the crack moving. If the geometry of the tip is retained during propagation the failure stress falls slowly as the crack length increases. If, however, the crack grows into a sharp crack the failure stress falls to the Griffith value.

Equation (9.19) may also be expressed in terms of Young's modulus E and the surface energy γ of the solid in the following way. The work done in pulling an ideally brittle solid specimen of unit cross-section apart is equivalent to the work done in creating two new unit surfaces, i.e. 2γ. As we have seen, this pulling apart occurs when the applied stress is of order $E/10$ [see equation (9.16)]. The separating atomic planes are pulled apart at this stage by a distance of the order $a/10$. The work done is of order $\dfrac{E}{10}\dfrac{a}{10} = \dfrac{Ea}{100}$ which can be equated to 2γ. Hence γ is of order $Ea/200$. Equation (9.19) then becomes

$$\sigma_g = \frac{E}{10} \sqrt{\left(\frac{3 \times 200\gamma}{El}\right)} = \sqrt{\left(\frac{600}{100} \times \frac{\gamma E}{l}\right)}.$$

A more rigorous analysis shows that the numerical factor under the square root sign is nearer unity than 6. The Griffith stress then becomes

$$\sigma_g \simeq \sqrt{\left(\frac{\gamma E}{l}\right)}. \tag{9.20}$$

The equilibrium spreading of a crack thus provides a means of determining the surface energy of a solid. We have already described one example of this in the splitting of a mica sheet. Attempts have also been made to apply this method to metals, polymers and other solids. Generally such experiments give extremely high values of the order of 100,000 erg cm.$^{-2}$; this is because plastic deformation occurs at the tip of the crack and the plastic work involved completely swamps the much smaller true surface energy. Recently Gilman has shown that with some ionic crystals it is possible to 'freeze-out' the ductility by carrying out the experiments with the specimens immersed in liquid nitrogen. Under these conditions surface energies of the order of a few hundred erg cm.$^{-2}$ are obtained.

Edge dislocations are too small to act as effective stress-raisers from the point of view of crack propagation. However, if the free movement of edge dislocations is obstructed by some barrier so that several can pile up in a row [see figure 59(c)], an internal crack may be formed; this may be able to initiate crack propagation. In this way a ductile solid may become brittle.

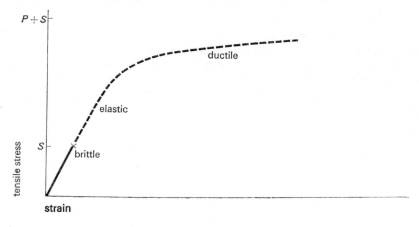

Figure 61. Stress–strain curve for a brittle solid. Brittle failure occurs for a tensile stress S. In the presence of a hydrostatic pressure P this is increased to $P+S$. The associated shear stress may produce plastic yielding before brittle failure occurs.

9·3·3   *How brittle solids may be made ductile*

Consider a brittle solid which fails in a brittle manner for a tensile stress S. If we apply a hydrostatic pressure the tensile stress necessary for failure is $P+S$. Associated with this tensile stress is a shear stress equal to $\frac{1}{2}(P+S)$. If the critical shear stress of the material is less than this it will flow in a ductile manner before the tensile stress is large enough to produce brittle failure. This is one of the reasons why rocks below the earth can flow in a ductile way although they are normally very brittle materials. Indeed Bridgman

169   The strength properties of solids

has shown in the laboratory that under sufficiently high hydrostatic pressure even quartz can flow plastically. In this connexion it is interesting to note that in indentation hardness experiments plastic indentation can often be made in relatively brittle materials (even though some cracking may also occur). This is because the large hydrostatic component of the stress field inhibits brittle failure. Furthermore the hardness values so obtained are a measure of the *plastic* properties of the brittle solid.

9·4 Conclusion

In this chapter we have discussed the strength and deformation properties of solids in terms of interatomic forces. The deformation characteristics fall into three main categories. When the stresses are below a certain level the strains are reversible and the deformation is said to be elastic. As we saw in the previous chapter elastic deformation involves the gentle distortion of the lattice and when calculations are possible there is good agreement between the observed behaviour and that derived from the interatomic forces. For larger stresses two other types of deformation occur, plastic and brittle: both are irreversible. These strength properties may also be calculated in terms of interatomic forces but observed values are generally very much smaller. We may say that solids owe their plastic and brittle strength to interatomic forces: their weakness to the presence of flaws or imperfections. The main characteristics are summarized in the following table.

Table 22 *Plastic and brittle failure*

Mode	Criterion	Source of weakness	Effect of hydro-static pressure
Plastic	Critical shear stress for planes to slide over one another	Dislocations, enable slip to occur row by row of atoms	None
Brittle	Tensile stress across a plane pulling planes of atoms apart	Cracks, act as stress-multipliers so that stress at crack tip is much greater than the applied stress	Opposes applied tensile stress and so can inhibit brittle failure

Chapter 10
Thermal and electrical properties of solids

In this chapter we shall discuss in terms of atomic mechanisms the specific heat of solids, thermal expansion and thermal and electrical conductivity.

10·1 Specific heat

10·1·1 *Definition of specific heat*

The specific heat or heat capacity is defined as the quantity of heat required to raise the temperature of one gram of substance by 1°C. With solids the specific heat at constant volume C_V is a little less than that at constant pressure C_P. For a gram-molecule the corresponding quantities are related by the thermodynamic function

$$C_P - C_V = \frac{\beta^2 T V}{K}, \tag{10.1}$$

where β is the volume coefficient of expansion, K the bulk modulus and V the volume of one gram-molecule. In practice the specific heat generally measured with solids is C_P, but the difference between this and C_V is small and can as a first approximation be neglected.

If we assume that the specific heat is due to the change in internal energy we may easily calculate it, say, for a gram-atom of a metal. This contains N_0 atoms. The atoms have no translational energy, only vibrational. Each vibrational degree of freedom involves both potential and kinetic energy so that the average thermal energy is kT per degree of vibrational freedom. Each atom has three independent degrees of vibration so that its average vibrational energy is $3kT$. For all N_0 atoms this gives a value of

$$U = N_0 \times 3kT = 3RT. \tag{10.2}$$

If we keep the volume constant any additional heat goes solely in increasing the vibrational energy. Hence

$$C_V = \left(\frac{\partial U}{\partial T}\right)_V = 3R, \tag{10.3}$$

where C_V is the specific heat per gram-atom. Thus C_V should have a value of about 6 cal. per degree, a conclusion reached empirically by Dulong and Petit in 1819. If one deals with a compound, say NaCl, then one would expect each gram-ion to possess a specific heat of 6 cal. °C.$^{-1}$; so that for the mole the specific heat would be 12 cal. °C.$^{-1}$; for a triatomic molecule it would be 18 cal. °C.$^{-1}$. Some typical results for C_P at room temperature for metals and diatomic and triatomic solids are given in Table 23. Ignoring the difference between C_P and C_V, it is seen that the specific heats per gram-atom are all indeed of the order of 6 cal. °C.$^{-1}$ (except ice).

Table 23 *Specific heats of solids*

Substance	C_P cal. g.$^{-1}$ °C.$^{-1}$	Atomic or Molecular weight g.	Specific heat C_P g-atom^{-1} or mole^{-1} cal.°C.$^{-1}$
Al	0·22	27	6
Cu	0·09	63·5	5·7
Ag	0·054	108	5·8
Pb	0·03	207	6·2
NaCl	0·21	57·5	12·0 for 2 gram-atom
CaF$_2$	0·22	78·0	17·1 for 3 gram-atom
SiO$_2$	0·28	60·0	16·8 for 3 gram-atom
Ice	0·5	18	9·0 for 3 gram-atom

10·1·2   *The Einstein model*

Later measurements showed that the Dulong and Petit law holds only above a certain temperature. At lower temperatures C_P falls off markedly. For example with copper the specific heat per gram-atom falls off from 5·7 at room temperature to 5·1 at −100°C., to 2·7 at −200°C. and to 0·2 at −250°C. In the limit it approaches zero at the absolute zero.

The first simple explanation of this was due to Einstein in 1906 who made use of Planck's earlier discovery of the distribution of thermal energy in an oscillating system. If each atom behaves as an independent harmonic oscillator of frequency v, its average energy (ignoring zero-point energy) is exactly the same as that given in our discussion of the vibrational energy of a diatomic molecule.

$$u = \frac{hv}{\exp\left(\dfrac{hv}{kT}\right) - 1}.$$ (10.4)

For the N_0 atoms each of which has 3 independent degrees of vibrational freedom we obtain

$$U = 3N_0 u. \tag{10.5}$$

For high temperatures this reduces to $3N_0 kT$ in agreement with the Dulong and Petit law. For lower temperatures it diminishes and as T tends to zero, it also tends to zero. If an appropriate value of v is assumed the shape of the Einstein theoretical curve agrees reasonably well with experiment except towards $T = 0$ (see figure 62). In this region careful experiments show that

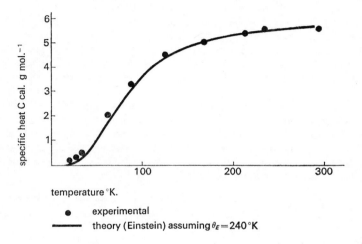

specific heat C cal. g mol.$^{-1}$

temperature $^\circ$K.

●   experimental

━━━   theory (Einstein) assuming $\theta_E = 240\,^\circ$K

Figure 62. Specific heat of copper as a function of temperature: ● experimental values, – theory given by Einstein assuming an atomic vibrational frequency of $v = 5 \times 10^{12}$ sec.$^{-1}$ If one writes $\dfrac{hv_E}{k\theta_E} = 1$ this gives for the Einstein temperature, corresponding to $v = 5 \times 10^{12}$ sec.$^{-1}$, a value of $\theta = 240^\circ$K.

C_V falls off as T^3, whereas in Einstein's theory it falls off as $\exp\left(-\dfrac{hv}{kT}\right)$, i.e., it falls off more rapidly.

10·1·3   *The Debye model*

There are two main defects in the Einstein model. First it assumes that all the atoms have the same single frequency v, and secondly that the vibration of each atom is independent of its neighbour. In fact the atoms act as coupled oscillators and a whole range of frequencies is possible. Each value of v (at a fixed temperature) contributes its own average thermal energy u [see equation (10.4)]. These should all be added to determine the total energy U of the whole solid. An analysis along these lines was first developed by Born and von

173   Thermal and electrical properties of solids

Karman in 1912. The main feature of such a treatment, which distinguishes it from the Einstein model, is that it reveals a whole spectrum of possible frequencies. The detailed analysis is, however, extremely complicated.

A completely different approach is that due to Debye (1912). The solid is treated as a continuum which, at first sight, seems a retrogressive step. However, its great merit is that it provides a frequency spectrum which is a close approximation to the true spectrum. Debye considers the way in which waves can travel through the solid. The most remarkable feature is that a *standing wave* of frequency v behaves exactly like a quantum oscillator of frequency v so that its thermal energy is again given by equation (10.4). This is surprising since the wave involves the vibration of all the atoms in the solid; yet its thermal energy is that of a *single* quantum oscillator. At first sight it would seem that this could not possibly give the same total energy. The difficulty, however, disappears when it is realized that the total vibration of each atom is the composite result of all the possible waves that can travel through the solid.

Suppose the solid is a crystal in the form of a cube of side L. To establish standing waves the free surface of the specimen must be either a node or an antinode and it may be shown that either condition leads to the same result. We shall treat the free surface as an antinode. Then the largest standing wave possible for a wave travelling normal to the faces of the cube has a wavelength $\lambda = 2L$ (see figure 63). If the velocity of the wave is C the frequency of the wave is $v = \dfrac{C}{\lambda} = \dfrac{C}{2L}$. The next possible standing wave has a wavelength

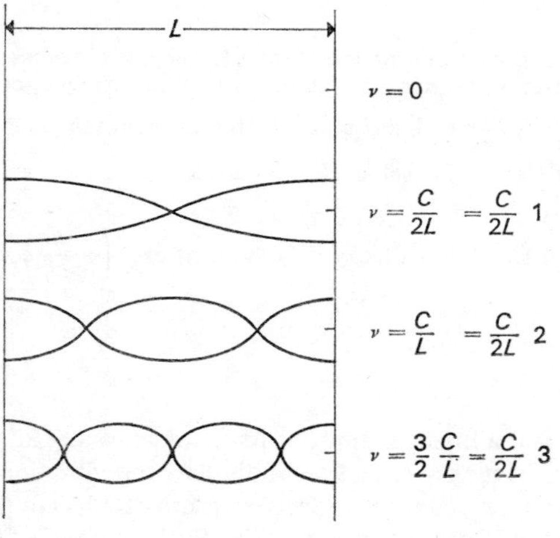

$v = 0$

$v = \dfrac{C}{2L} = \dfrac{C}{2L}\ 1$

$v = \dfrac{C}{L} = \dfrac{C}{2L}\ 2$

$v = \dfrac{3}{2}\dfrac{C}{L} = \dfrac{C}{2L}\ 3$

Figure 63. Characteristic frequencies of standing waves in an elastic continuum of length L. We assume that the free ends are antinodes and that the velocity C of the waves is independent of frequency.

$\lambda = L$ and its frequency is $v = C/L$. The next has a wavelength $\lambda = \frac{2}{3}L$ and its frequency is $v = \frac{3}{2} \times \frac{C}{L}$. Thus the possible frequencies increase in the order:

$$\frac{C}{2L} ; \frac{C}{2L} \times 2; \quad \frac{C}{2L} \times 3; \quad \frac{C}{2L} \times 4, \text{ etc. or } v_1, 2v_1, 3v_1, 4v_1. \tag{10.6}$$

The corresponding thermal energies are

$$\frac{hv_1}{\exp\left(\dfrac{hv_1}{kT}\right) - 1}, \quad \frac{2hv_1}{\exp\left(\dfrac{2hv_1}{kT}\right) - 1}, \quad \frac{3hv_1}{\exp\left(\dfrac{3hv_1}{kT}\right) - 1}, \text{ etc.} \tag{10.7}$$

and all that is now needed is to sum them. Very soon the possible frequencies become such large multiples of v_1 that they become virtually continuous and the sum can be replaced by an integral. One needs to know the number of vibrations that are possible between v and $v + dv$. This is not difficult for a continuum. A simple approach is as follows.

For the waves discussed above

$$v_n = \frac{C}{2L} n,$$

where n is an integer. If we considered the more general case of a wave travelling in some arbitrary direction we would obtain the same result but n would have components n_1, n_2, n_3 (each of them integers) in the x, y, z directions.

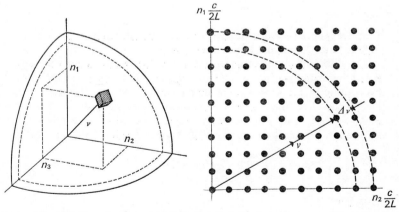

Figure 64. (a) Construction showing that the number of waves dN possessing frequency between v and $v + dv$ is proportional to the volume in the octet of the spherical shell lying between radii v and $v + dv$. (b) Two-dimensional figure showing that each point corresponding to a particular value of v occupies a cube of side $(C/2L)$.

175   Thermal and electrical properties of solids

Then

$$v = \frac{C}{2L} (n_1^2 + n_2^2 + n_3^2)^{\frac{1}{2}}. \tag{10.8}$$

We plot $n_1 \frac{C}{2L}$, $n_2 \frac{C}{2L}$, $n_3 \frac{C}{2L}$ on x, y, z co-ordinates. Then the distance from the origin to any point is equal to v. Since n_1, n_2, n_3 can only change by one unit at a time each possible point for v occupies a volume of $\left(\frac{C}{2L}\right)^3$. The number of points available for frequencies between v and $v+dv$ is the volume of the spherical shell contained between radii v, $v+dv$ divided by the unit volume $\left(\frac{C}{2L}\right)^3$. Since v can have only positive values we can only use one-eighth of the spherical shell. The resulting number is therefore

$$d\mathcal{N} = (\tfrac{1}{8} \times 4\pi v^2 dv) \left(\frac{2L}{C}\right)^3$$

$$= \frac{4\pi L^3 v^2 dv}{C^3}. \tag{10.9}$$

We allow for two transverse vibrations and one longitudinal vibration. If both have the same wave velocity C (or if we choose a suitable average velocity) this triples the value of $d\mathcal{N}$. Then the total thermal energy is simply

$$U = \int_0^{v_m} \left[\frac{hv}{\exp\left(\frac{hv}{kT}\right) - 1} \right] 3 d\mathcal{N}. \tag{10.10}$$

We have ignored zero-point energy since it is independent of T and when we differentiate U to find C_V it will disappear.

The difficulty about equation **(10.10)** is that we have no way, so far, of specifying the upper limit v_m of the integral. It is at this point that Debye ties his continuum model to a particulate model. If the crystal contains N atoms, Debye postulates that the total number of vibrations possible must be equal to $3N$ so that, in the limit, at high temperatures where each u has a value of kT the total thermal energy is $3NkT$. From equation **(10.9)** this means that

$$\int_0^N 3 d\mathcal{N} = \int_0^{v_m} 12\pi \frac{L^3}{C^3} v^2 dv = 3N. \tag{10.11}$$

This gives

$$v_m^3 = \frac{3}{4\pi} \frac{C^3}{L^3} N. \tag{10.12}$$

Before we use this we may form some estimate of its magnitude. If the solid has a simple cubic structure with a distance a between each atom, Na^3 will be the volume of the solid, i.e., L^3. Then

$$v_m = \left(\frac{3}{4}\right)^{\frac{1}{3}} \frac{C}{a} \simeq \frac{C}{a}. \tag{10.13}$$

This corresponds to a minimum wavelength

$$\lambda_{min} = \frac{C}{v_{max}} \simeq a. \tag{10.14}$$

Thus the limiting frequency on the Debye model corresponds to very short waves, the wavelength being comparable with the atomic spacing. This is reasonable since once we recognize the particulate nature of the solid, a wavelength less than the atomic spacing ceases to be meaningful. This result also corresponds rather closely to the very simple elastic model discussed in the previous chapter. We found that if the atoms have a natural (uncoupled) frequency v the velocity of a wave is of order $2\pi v a$ compared with $v_m a$ given in equation (10.13). Inserting equation (10.12) in (10.9),

$$3d\mathcal{N} = \frac{9N}{v_m^3} v^2 dv. \tag{10.15}$$

If this is substituted in equation (10.10) we have an explicit expression for U which can be evaluated. By differentiating, the specific heat C_V may be found.

The Debye theory predicts that the heat capacity of a solid depends only

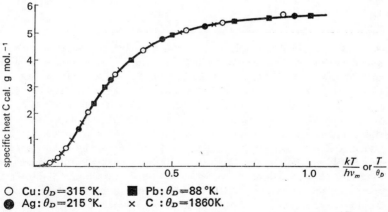

O Cu: $\theta_D = 315$ °K.   ■ Pb: $\theta_D = 88$ °K.
● Ag: $\theta_D = 215$ °K.   × C : $\theta_D = 1860$K.

Figure 65.   Specific heat of various solids plotted as a function of $\dfrac{kT}{hv_m}$ where v_m is the characteristic Debye frequency. Writing $\dfrac{hv_m}{k} = \theta_D$, the quantity θ_D is known as the Debye temperature.
o ● ■ × experimental points for Cu, Ag, Pb, diamond .———— theory given by Debye.

177   Thermal and electrical properties of solids

on the characteristic frequency v_m. This implies that if C_V for various solids is plotted against $\dfrac{kT}{hv_m}$ they should all lie on a single curve. This is found to be very nearly true (see figure 65). The quantity $\dfrac{hv_m}{k}$ is known as the Debye temperature θ_D. It is interesting to compare θ_D as calculated from equation (10.12) with the value of θ_D that gives the best fit of the Debye equation with the observed specific heat measurements.

Table 24 *The Debye temperature*

| | θ_D | |
Substance	From specific heat	From equation (10.12)
Pb	88	75
Cd	168	174
Ag	215	220
Cu	315	341
NaCl	281	305
CaF$_2$	474	510
Ice	220	200

Although the theory is only really valid for isotropic solids containing a single type of atom it is seen that it holds extremely well for more complicated materials. It even holds approximately for ice, which is rather surprising. (An average wave velocity of $2 \cdot 2 \times 10^5$ cm. sec.$^{-1}$ has been used to calculate θ_D). The specific heat of ice at 200°K is about 0·35 cal. g.$^{-1}$ or 6 cal. mole^{-1}. It thus behaves as a monatomic solid, practically the whole of the thermal energy being due to oscillations of the molecule as a whole. Between 200°K and 273°K (the melting point) there is some increase in specific heat to a value of about 9 cal. mole^{-1}. Presumably in this temperature range rotational degrees of freedom of the whole molecule contribute to the thermal energy of the ice crystal. As we shall see in Chapter 13 this rotational freedom accounts for the rather large low-frequency dielectric constant of well ice below 0°C (see figure 97).

Although the Einstein model is in many ways a simpler and more direct one than the Debye model there are two features in the Debye treatment that make it preferable. First v_m may be derived directly from bulk properties so that there are no assumed constants in the final calculation of U. By contrast the Einstein frequency (which turns out to be comparable with Debye's v_m, in fact for many solids, $v_E \simeq 0.75v_m$ or $\theta_E \simeq 0.75\theta_D$) must be deduced empirically from the shape of the C_V–T curve. Secondly at low temperatures the Debye relation for the specific heat reduces to

$$C_V = \frac{12\pi^4 R}{5\theta_D^3} \times T^3, \tag{10.16}$$

so that it satisfactorily explains the observed T^3 dependence in this temperature range.

The main defect of the Debye treatment is that it considers waves of all frequencies to travel at the same speed. This is not true of coupled oscillators. For higher frequencies there is an appreciable drop in wave velocity. Fortunately the waves in this range generally contribute only a small part to the total vibrational energy.

10·2 Thermal expansion: Gruneisen's law

We may now derive a simple relation between thermal expansion and specific heat. It is based on the existence of an asymmetrical potential energy curve for the whole crystal in terms of the separation between atoms. For simplicity we consider here only the potential energy u between one atom and its neighbour and ignore the problem of coupled interactions.

The potential energy curve is shown in figure 66(a); we transpose it to axes

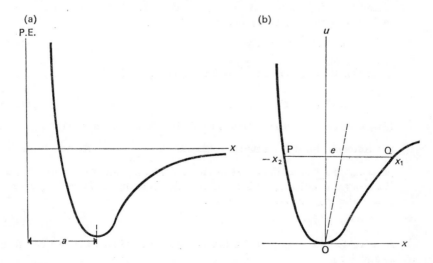

Figure 66. (a) Potential energy u of an atom as a function of its separation x from its neighbours, (b) same curve transposed to different axes so that the minimum has coordinates 0, 0.

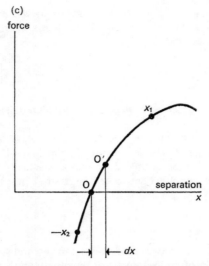

(c)

force

x_1

O'

O

separation
x

$-x_2$

dx

Figure 66. (c) force–separation curve for the situation described in (b); the expansion e is equivalent to an increase dx in the mean spacing.

as shown in figure 66(b) and express u in terms of the displacement x of the atoms from their equilibrium separation a. We assume that at 0°K. the energy is zero (ignoring zero-point energy) and that the addition of thermal energy increases u according to a power law:

$$u = Ax + Bx^2 + Cx^3 \dots .$$ (10.17)

Since O is the origin where $\dfrac{\partial u}{\partial x} = 0$ we obtain $A = 0$. Then

$$u = Bx^2 + Cx^3 + \dots .$$ (10.18)

If we used only $u = Bx^2$ the curve would be symmetrical about the u axis and the restoring force for small displacements, $-\dfrac{\partial u}{\partial x} = -2Bx$, would be proportional to x, so that the atoms would oscillate with simple harmonic motion. We now assume that the cubic term is sufficient to match the real asymmetric potential energy curve.

$$u = Bx^2 + Cx^3.$$ (10.19)

At a value of energy u_1 there are two possible values of x; we call them x_1 and $-x_2$.

$$u_1 = Bx_1^2 + Cx_1^3 = Bx_2^2 - Cx_2^3$$
$$B(x_1^2 - x_2^2) = -C(x_1^3 + x_2^3).$$ (10.20)

When adding x_1 and x_2 they can be considered as being almost equal. Equation (10.20) becomes

$$B(x_1 - x_2) = -Cx^2.$$ (10.21)

Oscillations between P and Q are not strictly simple harmonic because of the cubic term in x. The mean position, however, is approximately the mid-point of PQ. Its distance e from the u axis is $\frac{1}{2}(x_1 - x_2)$.

Hence, from (10.21)

$$e = \frac{-C}{2B} x^2.$$ (10.22)

The quantity e is the thermal expansion of the lattice when the mean vibrational amplitude is x. Its variation with temperature is

$$\frac{de}{dT} = \frac{-C}{2B} \frac{d}{dT} (x^2).$$ (10.23)

Now the thermal energy of vibration in one dimension is approximately $u = Bx^2$ so that the specific heat for a one-dimensional vibration is $\frac{\partial u}{\partial T} = B \frac{d}{dT} (x^2)$. Since this involves only the vibrational energy, it is the specific heat for constant volume. This makes no assumptions about equipartition of thermal energy so that for this model it holds however far the material may be from the Debye temperature. The specific heat per atom in three dimensions, c_V, is three times this, so that

$$c_V = 3B \frac{d}{dT} (x^2).$$ (10.24)

Hence from (10.23)

$$\frac{de}{dT} = \frac{-C}{6B^2} c_V.$$ (10.25)

The coefficient of linear expansion α is given by

$$\alpha = \frac{1}{a} \frac{de}{dT} = \frac{-C}{6B^2} \times \frac{1}{a} c_V.$$ (10.26)

Since a varies little with temperature we see that α is directly proportional to the specific heat at that temperature. The greater the specific heat, the greater the thermal vibration, and the greater the effect of the cubic term in augmenting the expansion. This is the main feature of the Gruneisen treatment.

It is possible to express B and C in terms of other bulk properties of the solid. For example, for a simple cubic array B is equal to $Ea/2$, where E is Young's modulus. The other parameter chosen by Gruneisen is the rate of change of vibrational frequency with lattice expansion. The natural frequency

181   Thermal and electrical properties of solids

of vibration v is proportional to (force constant)$^{\frac{1}{2}}$, i.e. to $\left(\dfrac{\partial^2 u}{\partial x^2}\right)^{\frac{1}{2}}$. Hence

$$\log v = \text{constant} + \tfrac{1}{2}\log\frac{\partial^2 u}{\partial x^2}.$$

$$\therefore \quad \frac{d}{dx}(\log v) = \frac{1}{2}\left(\frac{1}{\dfrac{\partial^2 u}{\partial x^2}}\right)\frac{\partial^3 u}{\partial x^3}. \tag{10.27}$$

From equation **(10.18)** we have

$$\frac{\partial^2 u}{dx^2} = 2B + 6Cx \simeq 2B; \qquad \frac{\partial^3 u}{dx^3} = 6C. \tag{10.28}$$

Hence

$$\frac{d}{dx}\log(v) = \frac{1}{2}\times\frac{6C}{2B} = \frac{3C}{2B}.$$

But $\qquad \dfrac{d}{dx}(\log a) = \dfrac{1}{a}$, $\tag{10.29}$

since on the force–displacement curve [figure 66(c)] the displacement of the equilibrium distance dx is the change in the spacing.

Consequently

$$-\frac{d\log v}{d\log a} = -\frac{3Ca}{2B}. \tag{10.30}$$

The Gruneisen constant γ is usually defined in terms of the change of frequency with atomic volume v. Since v is proportional to a^3 we have

$$\gamma = -\frac{d\log v}{d\log v} = -\frac{d\log v}{3d\log a} = -\frac{Ca}{2B}. \tag{10.31}$$

Substituting from equation **(10.31)** for $C = -\dfrac{2B\gamma}{a}$ and $B = \dfrac{Ea}{2}$ in equation

(10.26) we obtain

$$\alpha = \frac{2\gamma}{3Ea^3}\,c_V. \tag{10.32}$$

For a simple cubic structure $N_0 a^3$ is the volume V of a gram-atom and $N_0 c_V$ is the specific heat C_V of a gram-atom. Hence

$$\alpha = \frac{2\gamma}{3E}\times\frac{C_V}{V} \tag{10.33}$$

182   Gases, liquids and solids

(The coefficient of cubical expansion δ is 3α so that

$$\delta = \frac{2\gamma C_V}{EV}.$$

A more rigorous model gives

$$\delta = \frac{\gamma C_V}{KV},$$

where K is the bulk modulus.)

We see that α is expressed solely in terms of measurable bulk quantities. As a matter of interest we may calculate α for an ionic solid for which the potential energy between a pair of ions is

$$u = -k\left[\frac{1}{x} - \frac{a^8}{9x^9}\right]. \tag{10.34}$$

$$\ddot{u} = \frac{\partial^2 u}{\partial x^2} = -k\left[\frac{2}{x^3} - \frac{10a^8}{x^{11}}\right] = -k\frac{-8}{a^3} \quad \text{at } x = a,$$

$$\dddot{u} = \frac{\partial^3 u}{\partial x^3} = -k\left[\frac{-6}{x^4} + \frac{110a^8}{x^{12}}\right] = -k\frac{104}{a^4} \quad \text{at } x = a.$$

From equations (10.27) and (10.29) we see that

$$\gamma = -\frac{d \log v}{d \log v} = -\frac{d \log v}{3d \log a} = -\frac{a\,\dddot{u}}{6\,\ddot{u}}. \tag{10.35}$$

Consequently for an ionic solid

$$\gamma = -\frac{a}{6} \times \frac{104a^3}{8a^4} = -\frac{13}{6}.$$

For rock-salt $E \simeq 2 \cdot 10^{11}$ dyne cm.$^{-2}$; $\quad V \simeq 40$ cm^3 mole^{-1};

$C_V \simeq C_P \simeq 50 \cdot 10^7$ erg °C.$^{-1}$ mole^{-1}.

From (10.33) this gives

$$\alpha \simeq 10^{-4} \text{ °C.}^{-1}$$

The experimental value is of this order but smaller by a factor of about 4. With non-ionic solids γ is appreciably less and α is about 10 times smaller.

As it has been pointed out at the end of Chapter 7, the above treatment has one basic defect; namely that it treats the solid as though its behaviour were exactly analogous to that of a pair of atoms. By contrast when we consider the behaviour of an assembly of a large number of atoms as in a solid crystal, an entropy factor arises which is not accounted for by the simple treatment. It may indeed be shown that even if the atomic vibrations were perfectly harmonic, that is if the potential energy curve were perfectly symmetrical, the increase in free energy of the crystal as its temperature is raised would lead to an increase in volume [E. A. Guggenheim (1965), *Boltzmann's Distribution Law*, North-Holland Publishing Company]. However the thermal expansion due to this factor appears to be small compared with that arising from the asymmetric nature of the potential energy curve considered above.

10·3 Thermal conductivity

10·3·1 *Heat transfer*

There are three main forms of heat transfer.

(a) Radiation; this involves the emission of electromagnetic waves. All materials can radiate but radiation is generally most marked with solids.

(b) Convection; this involves the streaming of matter. With forced convection the streaming is imposed by some external driving mechanism; with natural convection it occurs as a result of density differences. Convection is observed only in liquids and gases.

(c) Conduction; this involves the transfer of thermal energy by some form of collision. It occurs in gases, liquids and solids. With poorly conducting solids the energy transfer is provided by lattice vibrations; with metals this is greatly augmented by electron collisions. In this chapter we shall deal solely with conduction in solids.

The rate at which heat flows through any element of a solid is proportional to the cross-sectional area dA of the element and the temperature gradient measured normal to dA. We may make this proportionality into an equality by inserting a constant K which we call the thermal conductivity. Then

$$\frac{dQ}{dt} = -K \, dA \times \frac{dT}{dx}. \tag{10.36}$$

The negative sign shows that a positive flow of heat takes place from a hotter to a cooler region, i.e., where dT/dx is negative. K has the dimensions $MLT^{-3}\theta^{-1}$. It varies from the best thermal insulators to the best conductors by a factor of only about 1000. By contrast, electrical conductivities can vary by a factor of 10^{30}.

10·3·2 *Heat flow down a uniform bar*

As a simple example we consider the heat flow down a uniform bar under two extreme conditions. When it is

(a) thermally insulated. In the steady state if no heat escapes from the surface of the bar the quantity of heat crossing every section per second must be constant. The temperature gradient must therefore be constant [see equation **(10.36)**]. Knowing the temperature of the hot and cold ends, the cross-section of the bar and the rate of heat conduction, K may be determined directly. This is a convenient method of measuring K for good conductors. It may also be used with poor conductors if the bar is replaced by a thin disc of the material.

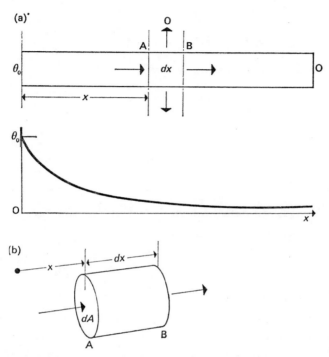

Figure 67. (*a*) Steady-state flow of heat along an exposed uniform bar. The heat flow in at A, less the heat flow out at B is equal to the heat lost by radiation and convection from the surface between A and B. (*b*) Heat flow through a thermally insulated element: the heat flow in at A, less the heat flow out at B is equal to the heat accumulated in the element.

(*b*) exposed to a cooler constant temperature environment. Consider a very long bar of uniform cross-sectional area A held at a temperature T_0 at one end (figure 67). Let the environment be at a temperature T_e. Consider a small element of length dx at some point along the bar where its temperature is T. Express these as temperature excesses over the environmental temperature. The hot end is at θ_0, the element at θ and the environment at 0°. In the steady state heat enters at A, some is lost by radiation and convection from the surface of the element, and the remainder passes through the element at B. If the circumference of the

185 Thermal and electrical properties of solids

bar is c the surface area of the element is cdx. If the heat lost is proportional to the surface area, the emissivity ε of the surface, and the temperature excess over the surroundings, the heat lost per sec. is $cdx\,\varepsilon\,\theta$.

Heat flow per sec. in at $A = -KA\,\dfrac{d\theta}{dx}$.

Heat flow per sec. out at $B = -KA\,\dfrac{d}{dx}(\theta+d\theta) = -KA\left(\dfrac{d\theta}{dx}+\dfrac{d^2\theta}{dx^2}\,dx\right)$.

Heat flow per sec. at A exceeds outflow at B by

$KA\,\dfrac{d^2\theta}{dx^2}\,dx$ = heat lost from surface of element.

$$KA\,\frac{d^2\theta}{dx^2} = c\varepsilon\theta. \tag{10.37}$$

The solution is of the form

$$\theta = Be^{wx}+Ce^{-wx},$$

where $w^2 = \dfrac{c\varepsilon}{KA}$. At great distances from the hot end the temperature approaches the ambient, i.e., $\theta = 0$. Hence $B = 0$

$$\theta = Ce^{-wx}. \tag{10.38}$$

Since at $x = 0$, $\theta = \theta_0$ we have finally

$$\theta = \theta_0 \exp\left[-\left(\frac{c\varepsilon}{KA}\right)^{\frac{1}{2}}x\right]. \tag{10.39}$$

The temperature falls off exponentially with distance from the hot end.

10·3·3   *Thermal diffusivity*

Consider a body which is insulated so that no heat is lost to the surroundings. We heat one part of it and consider how the heat flows before steady-state conditions are reached. Clearly the body has to be warmed up (i.e., it must absorb heat) as the heat flows through it.

Consider a small element of cross-section dA at a point A (ordinate x) where the temperature is T. At the neighbouring point B (ordinate $x+dx$) the temperature is $T+dT$ [see figure 67(b)].

Heat flow per sec. in at $A = -KdA\,\dfrac{dT}{dx}$.

Heat flow per sec. out at $B = -KdA\,\dfrac{d}{dx}(T+dT)$.

The heat accumulated in the element per sec. $= KdA \dfrac{\partial^2 T}{\partial x^2} dx$. If the element has density ρ, specific heat C, its heat capacity is $\rho\, dA\, dx\, C$. If the temperature rise per sec. of the element is $\dfrac{dT}{dt}$ we have

$$\rho\, C\, dA\, dx\, \frac{dT}{dt} = K dA \frac{\partial^2 T}{\partial x^2} dx, \tag{10.40}$$

$$\frac{dT}{dt} = \frac{K}{\rho C} \frac{\partial^2 T}{\partial x^2}.$$

The quantity $\dfrac{K}{\rho C}$ is known as the thermal diffusivity or *temperature* conductivity. It is of great importance in non-steady-state thermal problems.

10·3·4   *Theory of thermal conductivity*

The thermal conductivity of solids can be explained in terms of some type of collision process by means of which thermal energy is transported from the hotter to the colder regions. It is natural therefore to express it in a manner resembling the treatment for gases. For a gas where the transfer is due to molecular collisions we have

$$K = \tfrac{1}{3}nm\bar{c}\lambda C_V, \tag{10.41}$$

where C_V is the specific heat per unit mass. We note that the quantity nmC_V is the specific heat per unit volume s. Then

$$K = \tfrac{1}{3}s\bar{c}\lambda. \tag{10.42}$$

10·3·4·1   *Poor conductors.* In poor conductors the thermal energy is transferred by lattice vibrations. These travel through the specimen as waves which are scattered by the lattice as they progress through it. The wave is attenuated corresponding to the drop in temperature through the specimen. These thermal waves thus behave like particles possessing energy and momentum: they are known as *phonons*.

We may treat the waves as energy packets moving with the velocity c of a sound wave, and travelling a distance λ before they collide and communicate their thermal energy to the lattice. By exact analogy with gases we may write

$$K = \tfrac{1}{3}(\text{specific heat per unit volume})\,(\text{velocity of sound wave})\lambda.$$

This relation is found to be quite satisfactory for most poor conductors, the value of λ being of the order of 20 Å.

10·3·4·2   *Good thermal conductors: metals.* Good thermal conductors are generally good electrical conductors – their thermal conductivity is 100 to 1000 times greater

than that of insulators. This suggests that the major part of the thermal conductivity, like that of the electrical conductivity, arises from movement of electrons.

We make the following assumptions about the metallic state:

(i) The metal consists of fixed positive ions in a sea of electrons: in general there will be one or two electrons per ion.

(ii) The electrons behave as a perfect gas transporting thermal energy as in gas conduction.

(iii) Each electron has thermal energy of translation equal to $\frac{3}{2}kT$, i.e., its thermal capacity is $\frac{3}{2}k$.

(iv) Each electron travels a distance λ before colliding with a positive ion and giving up all its thermal energy.

We rewrite equation (10.41) in the form

$$K = \tfrac{1}{3}n\bar{c}\lambda(mC_V), \tag{10.42a}$$

and note that mC_V is the thermal capacity of a single travelling particle. Replacing this by $\frac{3}{2}k$ for our electron gas, we obtain

$$K = \tfrac{1}{2}n\bar{c}\lambda k, \tag{10.43}$$

where $\bar{c}$ is the velocity of an electron assuming it to behave as a gas molecule. At room temperature its value is of order 10^7 cm. sec.$^{-1}$ If we consider a monovalent metal such as sodium, for which there is a single electron per ion, we have $n \simeq 50 \times 10^{22}$, $k = 1\cdot4 \times 10^{-16}$. The observed conductivity at room temperature is about $1\cdot3 \times 10^7$ erg cm. °C. sec.$^{-1}$ This gives $\lambda \simeq 3$ Å. In general, equation (10.43) gives a good agreement with observation on the assumption that λ is of the order of the atomic spacing. Although this seems satisfactory there is a major defect in the theory which we shall discuss later.

10·4 Electrical conductivity of metals

10·4·1 *Drude model*

The general picture that we adopt here, which is due to Drude (1912), is similar to that used in the preceding section. It is assumed that the electrons move like a perfect gas in all directions through the lattice and that their thermal motion is terminated by collisions with positive ions. Under normal conditions there is no net transfer of charge in any direction. When, however, a potential is applied across the metal the electric field imposes a drift velocity on the electrons and it is this which is responsible for the observed electrical conductance. The drift velocity is very small compared with the mean electron-gas velocity $\bar{c}$.

Consider a bar of uniform cross-section A, length L, across which a potential

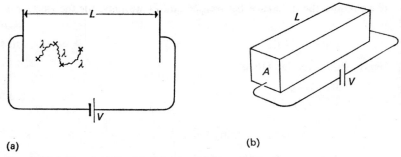

Figure 68. Electrical conductivity model for metallic conductors.

V is applied [figure 68(b)]. The electric field on the electrons is $X = \dfrac{V}{L}$. The force on an electron is eX and its acceleration is $\alpha = \dfrac{eX}{m}$. If λ is the mean free path of the electron the time between collisions is $t = \lambda/\bar{c}$. The drift velocity acquired during this period is αt so that the average drift velocity u between collisions is $\frac{1}{2}\alpha t$.

$$u = \tfrac{1}{2}\alpha t = \frac{1}{2}\frac{eX}{m}\frac{\lambda}{\bar{c}}. \qquad (10.44)$$

The model still assumes that the electron gives up the velocity gained from the field by colliding with a positive ion. If n is the number of free electrons per c.c. the number travelling per sec. along the bar due to drift is the number in a volume Au, i.e., Aun. Each electron carries a charge e so that the charge carried per sec., i.e., the current i, is

$$i = eAun$$

$$= \frac{1}{2}\frac{AX\lambda e^2 n}{m\bar{c}}, \qquad (10.45)$$

or $\qquad i = \frac{AV}{L}\times\frac{\lambda e^2 n}{2m\bar{c}}. \qquad (10.46)$

If ρ is the specific resistivity and κ the specific conductivity we know that

$$i = \frac{AV}{L}\times\frac{1}{\rho} \quad\text{or}\quad \frac{AV}{L}\times\kappa. \qquad (10.47)$$

Comparison with (10.46) shows that

$$\kappa = \frac{\lambda e^2 n}{2m\bar{c}} = \frac{\lambda e^2 n\bar{c}}{2m\bar{c}^2}. \qquad (10.48)$$

If we ignore the difference between $\bar{c}^2$ and $\overline{c^2}$ we may, for the electron gas, write

$$\tfrac{1}{2}m\bar{c}^2 = \tfrac{3}{2}kT. \tag{10.49}$$

Hence, from equation (10.48)

$$\kappa = \frac{\lambda e^2 n\bar{c}}{6kT}. \tag{10.50}$$

For a monovalent metal assuming $\lambda = 3$ Å., $n = \dfrac{1}{27 \times 10^{-24}}$, $e = 1\cdot6 \times 10^{-20}$

e.m.u., $\bar{c} = 10^7$ cm. sec.$^{-1}$, $k = 1\cdot4 \times 10^{-16}$, $T = 300°$K.,

$$\kappa \simeq 10^{-5} \text{ e.m.u.} \tag{10.51}$$

This is in reasonable agreement with the resistivity values of most metals, $\rho \simeq (2 \text{ to } 20) \times 10^{-6}$ ohm cm., or conductivity $= 5 \times 10^{-4}$ to 5×10^{-5} e.m.u.

10·4·2   *Wiedemann–Franz relation*

We may combine equations (10.43) and (10.50) to find the ratio of thermal to electrical conductivity. We have

$$\frac{K}{\kappa} = 3\left(\frac{k}{e}\right)^2 T. \tag{10.52}$$

Historically, Wiedemann and Franz observed that K/κ is the same for all metals at the same temperature. This was extended by Lorenz who observed that the ratio was proportional to the absolute temperature. The above analysis contains both 'laws'. The theoretical ratio is indeed close to the observed values over an appreciable temperature range for a large number of metals.

10·4·3   *Specific heat of the free electron*

In the Drude model discussed above the electron gas is assumed to possess thermal energy. Like any monatomic gas it should therefore contribute a specific heat of $\tfrac{3}{2}R$. The specific heat of a monovalent metal should therefore consist of $3R$ from its vibrational (lattice) energy and $\tfrac{3}{2}R$ from its free electrons, i.e., the specific heat should have a value of about 9 cal. g-atom^{-1} °C.$^{-1}$ The observed value in fact does not exceed 6.

This is a basic defect of the electron gas model and the explanation can only be given in terms of the energy states of the electrons in a metal. No energy state can contain more than two electrons (of opposite spins), so that the electrons fill the available states from the lowest available energy level until they reach the topmost or Fermi level, E_F (see figure 69). This is the situation at absolute zero. If the metal is heated only those electrons, which can be moved into a higher vacant state above E_F, can acquire further energy. In effect this means that only the electrons at the top of the Fermi level can acquire energy. Analysis shows that if there are n free electrons per c.c. the number of thermally excited electrons at temperature T is

$$\Delta n \simeq n \times \frac{kT}{E_F}.$$

(10.53)

For a typical metal $E_F \simeq 2$ eV. and at room temperature ($T \simeq 300°$K.) kT is about 4×10^{-14} erg $\simeq 0.025$ eV. Thus Δn is only about 1 per cent of n. If the thermal energy acquired by these electrons is of order kT we see that the specific heat is only about 1 per cent of the value quoted above. It is for this reason that the electronic contribution to the specific heat is so small.

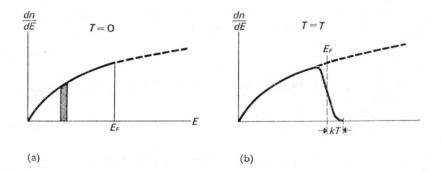

Figure 69. Electronic energy levels in a metal shown as $\frac{dn}{dE}$ plotted against the energy E so that the shaded area represents the number of electrons per c.c. with energy between E and $E + dE$. (*a*) At absolute zero all the energy states are filled up to a level E_F known as the Fermi level. (*b*) At a higher temperature T a small number of electrons at the top of the Fermi level are able to acquire additional energy of order kT.

We may take this a stage further. Since the specific heat of a single electron is of the order k, we see from equation (10.53) that the specific heat per c.c. is of order

$$k\Delta n = \frac{nk^2 T}{E_F},$$

(10.54)

so we may regard the *average* specific heat per electron as

$$c_e = \frac{k^2 T}{E_F}.$$

(10.55)

The expression from equation **(10.42a)** for the thermal conductivity may be written

$$K = \tfrac{1}{3}n\bar{c}\lambda(c_e),$$ (10.56)

whilst the electrical conductivity, from equation **(10.48)**, may be written

$$\kappa = \frac{\lambda e^2 n}{2m\bar{c}}.$$ (10.57)

The ratio is

$$\frac{K}{\kappa} = \frac{2}{3}\frac{m\bar{c}^2}{e^2}(c_e).$$ (10.58)

(At this point if we assume the electrons behave like a perfect gas $m\bar{c}^2$ is $3kT$ and $c_e = \tfrac{3}{2}k$; we then obtain precisely the classical Drude result given in equation **(10.52)**.)

The quantity $\bar{c}$, the velocity of those electrons in the Fermi distribution which take part in the thermal transport process, is of order

$$\bar{c}^2 \simeq \frac{E_F}{m},$$ (10.59)

so that equation **(10.58)** becomes, after substituting for c_e from equation **(10.55)**,

$$\frac{K}{\kappa} \simeq \frac{E_F}{e^2}(c_e) \simeq \frac{k^2 T}{e^2}.$$ (10.60)

A more rigorous treatment gives a constant $\left(\dfrac{\pi^2}{3}\right)$ in front of this expression so that it is very nearly the same as equation **(10.52)** obtained from the simple Drude model.

The conclusions one derives from this, however, expose another issue of very great importance. The value of E_F of about 2 eV implies that the electrons have velocities [see equation **(10.59)**] about 10 times greater than the free electron-gas velocity. To explain the observed conductivities one must assume a mean free path λ of the order of a few 100 Å. On the Drude model λ has a value comparable to the distance between the positive ions in the lattice; this seems reasonable. A value of several 100 Å. seems strange. The explanation lies in the fact that electrons travelling through a perfect lattice experience no resistance whatsoever. If we treat the electrons as waves it is possible to show that electrons scattered by the lattice recombine to give a wave of undiminished amplitude; in other words there is no attenuation of the wave so that the resistance is zero. At normal temperatures the lattice vibrations render the lattice irregular so that the electron waves are scattered in an irregular way. It is this which accounts for the observed finite resistance. Any factor which increases the irregularity of the lattice increases the resistance. This is why the resistance increases with temperature and with impurity content.

Chapter 11
The liquid state

The main characteristics of the gaseous state are random, translational motion of the molecules and relatively little molecular interaction. In the solid state the main characteristics are fixed sites for the individual atoms or molecules and very little translational movement. Both these states are well understood. By contrast the liquid state, in some ways, has 'no right to exist'. The molecular interaction must be quite strong since a given quantity of liquid, like a solid (and unlike a gas) occupies a definite volume. On the other hand the molecules have so much freedom that a liquid, unlike a solid, flows very readily and easily takes up the shape of the vessel it occupies.

The gaseous state has been thoroughly investigated and most of its problems have been satisfactorily resolved. The solid state is still being studied in hundreds of physics laboratories all over the world. By contrast the liquid state is a neglected step-child of the physical scientists. This is partly because the liquid state raises a number of very difficult theoretical problems. It is also partly a matter of fashion dictated by the economic value and engineering uses of solids as structural materials, electrical devices, etc. But for chemists, physical chemists and especially biologists, the liquid state is possibly of greater importance than the solid state. We may well expect, we should certainly hope for, marked advances in our understanding of the liquid state in the next two or three decades.

There are basically three approaches to the liquid state. The first is to treat it as a modified gas; the second is to treat it as a modified solid; the third is to treat it *sui generis* – this is the most difficult of all.

11·1   The liquid as a modified gas

A liquid in static equilibrium cannot withstand a shear stress. If we compress a column of liquid with a piston the situation is not like that of a solid where a uniaxial pressure or tension produces a shear stress in the medium. Unlike a solid the liquid must be held in a container; the walls exert pressures on the liquid in such a way that no resultant shear stress remains. This is equivalent to saying that the only stress the liquid can experience is a hydrostatic pressure or tension. It is for this reason that, at any point in a stationary liquid, the pressure is the same in all directions. In this sense a liquid closely resembles a gas and indeed both states are sometimes designated by the single generic term 'fluid'. In terms of resistance to pressure it therefore seems reasonable to consider how far the gas laws can be applied to explain the behaviour of liquids. Naturally we would not expect the ideal gas laws to be applicable but we might expect the imperfect gas laws to have some relevance.

We start by considering a liquid as a limiting case of a van der Waals gas. We ask what meaning may be attached to the van der Waals constants in the equation

$$\left(P+\frac{a}{V^2}\right)(V-b) = RT.$$

(a) The quantity b. In a gas $b = 4v_m$ and at the critical volume $V_c = 3b$ so that the quantity (V_c-b) is still positive and meaningful. At temperatures well below the critical temperature the density of the liquid increases, the volume occupied by the liquid is only perhaps $(1.5$ to $2)v_m$ so that $(V-b)$ is negative unless a different correction value is given to b.

(b) The quantity a/V^2. So long as the molecular volume is large a/V^2 is small and is meaningful as a correction term. However, when V is comparable with or less than b the term a/V^2 has a value of the order of 1000–10,000 atmospheres. This internal pressure is so large compared with the atmospheric pressure that it cannot be treated as a correction factor. This is, of course, consistent with the fact that the external pressure P is not an important factor in determining the volume occupied by a liquid.

(c) The tensile strength of a liquid. In what follows we shall show that for low-temperature isotherms a liquid, according to van der Waals' equation, can actually withstand negative pressures. If we apply an increasing negative pressure (hydrostatic tension) the liquid will expand until at some critical value vapour will begin to form within the liquid, that is to say the liquid begins to cavitate or pull apart. The pressure at which this occurs provides, therefore, a measure of the tensile strength of the liquid.

Following Cottrell (1965) we use the reduced equation of state and write

$$\left(\pi+\frac{3}{\phi^2}\right)(3\phi-1) = 8\theta, \tag{11.1}$$

where π, ϕ and θ are the reduced pressure, volume and temperature respectively. We may expand the liquid from D to B (see figure 70) by applying a hydrostatic tension. At B the liquid would become a vapour, that is it would pull apart. The pressure at which this occurs may be obtained by writing equation (11.1) as

$$\pi = \frac{8\theta}{(3\phi-1)}-\frac{3}{\phi^2}, \tag{11.2}$$

and determining the minimum at B by putting $\frac{\partial \pi}{\partial \phi} = 0$. This gives

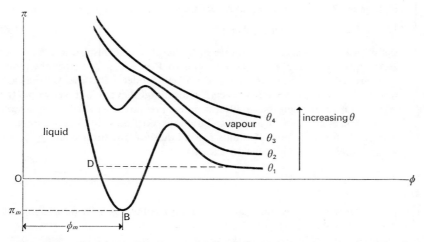

Figure 70. The tensile strength of a liquid treated as an extreme case of a van der Waals fluid. Using the reduced equation of state it is shown that the liquid fails when a negative pressure π_m is applied.

$$\frac{-3 \cdot 8\theta}{(3\phi-1)^2} + \left(\frac{6}{\phi^3}\right) = 0.$$

This defines the value of ϕ at **B**; we call this ϕ_m. Then

$$4\phi_m^3\theta = (3\phi_m - 1)^2. \tag{11.3}$$

If we insert this in equation **(11.2)** we obtain the corresponding value of π which we call π_m.

$$\pi_m = \frac{3\phi_m - 2}{\phi_m^3}. \tag{11.4}$$

We see from this equation that $\pi_m = 0$ when $\phi_m = \frac{2}{3}$. This is the 'last' isotherm for which the liquid has no tensile strength; it occurs when $\phi = \frac{2}{3}$ or $V = \frac{2}{3}V_c$. If the liquid is less condensed than this it can exist only if a positive pressure is applied.

At lower temperatures $-\pi$ becomes larger. It reaches its maximum (negative) value, as equation **(11.3)** shows, when $\theta = 0$, that is when $3\phi = 1$. Putting $\phi = \frac{1}{3}$ in equation **(11.4)** we obtain as the maximum value

$$\pi_m = \frac{1-2}{(\frac{1}{3})^3} = -27. \tag{11.5}$$

The limiting hydrostatic tension is then

$$P = -27P_c.$$

195   The liquid state

For water $P_c = 218$ atmospheres, so that

$$P = -5600 \text{ atmospheres.} \tag{11.6}$$

Before we leave this calculation we note that $\theta = 0$ corresponds to absolute zero temperature and $\phi = \frac{1}{3}$ corresponds to $V = \frac{1}{3}V_c = b$, that is the molecules are packed as close as the van der Waals model will allow. We may ask how such high negative pressures can be meaningful in a liquid. Why does the vapour not form when the smallest negative pressure is applied? The answer is that the nucleation of a vapour bubble is very difficult. For example to open up a spherical hole of radius r in a liquid of free surface energy γ, the vapour pressure would have to exceed

$$p = \frac{2\gamma}{r}. \tag{11.7}$$

For water $\gamma = 72$ erg cm.$^{-2}$ and assuming r is the size of a water molecule (= about 3 Å.) we obtain

$$p = 4800 \text{ atmospheres.} \tag{11.8}$$

Generally the vapour pressure is a negligible fraction of this value. However, we could produce the nucleation of bubbles, even if the vapour pressure is very small, by applying an external hydrostatic tension of 4800 atmospheres. Cavities would form and the liquid would then rupture. This tension is very close to that calculated in equation (11.6).

Readers may note that we could, in fact, estimate the breaking strength of the liquid simply in terms of the surface energy. Consider a liquid column of unit cross section. If we pull it apart we create two unit areas of surface, so that the work done is 2γ. The intermolecular forces are short range and if they become negligible for a distance greater than say x, the force F to pull the column apart will be of order $F \times x = 2\gamma$ or $F = 2\gamma/x$. If x is about 5 Å. we again obtain as the breaking stress (force per unit area) a value close to that given in equations (11.6) and (11.8). Exact agreement is not to be expected; the main point we wish to make is that because of intermolecular forces liquids can, in theory, withstand large negative pressures. In practice, of course, liquids are generally much weaker because of dissolved gases which readily nucleate cavities as the negative tension is applied. This is particularly marked at the walls of the vessel where small amounts of gas may be trapped in the surface irregularities. Carefully outgassed liquids in smooth-walled containers can, however, possess tensile strengths up to one tenth of the values calculated above.

11·2 The structure of liquids: the 'radial distribution function'

Before turning to a consideration of a liquid as a modified solid we mention one characteristic feature of a liquid which distinguishes it in a very basic way from a gas. A gas has no structure. If we study the diffraction of neutrons, electrons or X-rays by a monatomic gas we find that there is no coherent diffraction; the gas behaves like a diffraction grating in which the spacings between the lines are random. (For diatomic molecules one obtains diffraction maxima corresponding to the mean distance between the atoms in the molecule.) If, however, a monatomic gas is liquefied it shows marked coherent diffraction; the maxima are not sharp as with a solid crystal, but are diffuse.

The information obtained by such diffraction experiments is presented most graphically by means of the 'radial distribution function' which may be derived in the following way. We choose a molecule at random within the liquid and draw a series of concentric spheres around it such that the volume interval between two neighbouring spheres is a constant. The distribution function is the average number of molecules contained between each spherical shell. It is evident that the radial distribution function is a measure of the average density as a function of distance from some arbitrary origin. A typical result from X-ray diffraction is given in figure 71 and similar results are obtained from

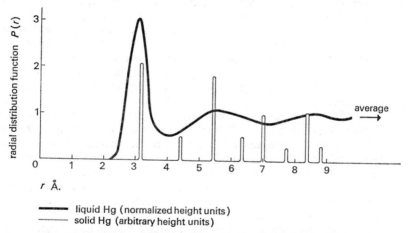

liquid Hg (normalized height units)
solid Hg (arbitrary height units)

Figure 71. Radial distribution function plotted against radius measured from an arbitrary atom chosen as origin. (a) Liquid mercury (b) solid mercury – from results given by A. Guinier, *X-ray Diffraction in Crystals, Imperfect Crystals and Amorphous Bodies*, Freeman, San Francisco, 1963, p. 70, figure 3·9. For the liquid the vertical ordinate can be normalized such that for large distances the radial distribution function tends to unity (uniform density). For the solid this is not possible since in principle each peak is infinitely narrow and infinitely high. Instead the vertical ordinate has been drawn proportional to the number of atoms at the specified radial distance. It is seen that with the liquid the structure is lost beyond distances of 10Å; with the solid it persists indefinitely.

the diffraction of neutrons. As a matter of interest the radial distribution function for the crystalline solid is also shown; this brings out the marked

197 The liquid state

difference between the liquid and the solid state. The solid behaves like a diffraction grating containing a very large number of regularly spaced lines: the liquid resembles the diffraction of light by a grating containing only a few lines. This suggests that there is short-range order in a liquid; any small group of atoms or molecules is in fairly regular array and resembles a crystal lattice but it is just a little out of register in a random way so that, as soon as one moves a short distance from the group, the neighbouring atoms (or molecules) no longer possess the long-range order characteristic of the solid crystal.

The contrast between the liquid and solid states may be expressed in another way (see J. D. Bernal, *Scientific American*, no. 203, 1960, pp. 124–34, and *Proceedings of the Royal Society*, A280, 1962, pp. 299–322). Neutron diffraction experiments suggest that a molecule in a liquid has time to vibrate 10 to 100 times before the local structure changes. During that time the structure of a liquid is physically, though not geometrically, similar to that of a crystal. The difference, however, is crucial. For example, the solid crystal, because of its high degree of order, admits only a limited degree of variation of composition. Crystallization as a means of purifying substances is evidence of the particular architecture that marks each crystalline phase. By contrast, liquids mix much more readily either with other liquids or with solids or with gases (though of course they do not mix as readily as gases which are always perfectly miscible with one another). We may describe this situation by saying that 'atoms in a liquid do not occupy such closely specified positions as they do in solids. They have more elbow room, and they are not so particular about their partners.'

Nevertheless the structural resemblance to the solid over limited regions appears to be indisputable. This conclusion is supported from a completely different type of measurement. It is found that when metals melt their electrical resistance increases by a factor of only about two. As mentioned in a previous chapter the resistance is primarily due to non-coherent scattering of the electrons by the metal ions. This implies that, to the conductance electrons, the metal ions, in the liquid, appear to be oscillating about regular sites, very much as in the solid state.

There are two main differences between the solid and liquid state. First, as discussed above, there is the absence of long-range order in liquids: in this sense they closely resemble amorphous solids, glasses and polymers. It is significant that these materials can flow viscously like liquids, although as pointed out in Chapter 7 their viscosity is generally enormous compared with that of true liquids. On the other hand, with crystalline solids flow or 'creep' can occur, but it takes place extremely slowly and through an entirely different mechanism – through the movement of dislocations, the behaviour of which is determined by the highly ordered structure of the material. This brings us to the second major difference: compared with the molecules in the solid state (both crystalline and amorphous), the molecules in liquids have very much greater mobility. They can swop places or shuffle around to other positions with relative ease.

11·3 The liquid as a modified solid

When solids melt they expand generally by 5–15 per cent in volume. One way of describing this is to say that there is an increase in free volume between the constituent molecules or atoms. Another is as follows (cf. E. A. Moelwyn-Hughes, private communication; see also *Physical Chemistry*, Pergamon Press, 1960). If one thinks of a typical crystalline solid for which each molecule, or atom, has say 8 or 12 nearest neighbours, we may regard the liquid state as a solid from which one nearest neighbour has been subtracted. This leaves the material with the right amount of volume expansion and a reasonable degree of short-range order. What is the nature of this disorder? The X-ray diffraction evidence suggests that the mean intermolecular distance does not change as much as would be expected from the bulk volume expansion. For example, when solid argon melts the mean argon–argon spacing increases by only about 1 per cent, whereas the volume change would lead us to expect a mean increase in separation of about 5 per cent. Such evidence may be used to justify the view that the melting process consists mainly in producing 'holes' in a structure which otherwise closely resembles the solid structure. This idea was first developed by J. G. Kirkwood and constitutes the simplest form of the 'hole-theory' of liquids. As we shall see later the modified 'hole-theory' introduced by H. Eyring provides a very convenient model for the study of viscous flow in liquids.

Attempts have been made to treat the disordered structure and the increase in free volume in the liquid state in terms familiar to the solid-state physicist. One approach is to concentrate on the role of vacancies or holes (A. L. G. Rees, 1966, private communication). As the temperature rises the concentration of vacancies increases. So long as vacancies are individual holes the behaviour is still that of a solid. As soon, however, as two or three vacancies coalesce a whole aggregate of molecules surrounding the hole becomes relatively mobile. The material is now very much more like a liquid. One advantage of this approach is that it stresses the co-operative nature of the melting process.

11·4 The liquid state *sui generis*

The most difficult approach is to tackle the liquid state without directly referring to the gaseous or solid state. The simplest method is that due to J. D. Bernal and his colleagues in England and to G. D. Scott in Canada. It may be referred to as a random packing model. Recognizing the disordered but relatively close-packed nature of the liquid structure Bernal took a large number of plasticine spheres, chalked them to stop them sticking, placed them in a football bladder, removed the air to avoid bubbles and then squeezed them together until they filled the whole of the space. On examining the aggregate it is found that the spheres have become polyhedra of various irregular shapes. The most common number of faces is thirteen (this is not the true co-ordination number since the centres of some of the neighbours are more than the average distance apart). More striking is the observation that the most common number of sides to a face is five. This is of great significance for the following reason. In crystallography molecules can be arranged with two-fold, three-fold, four-fold, or six-fold symmetry, never with five-fold symmetry. This is because 'one cannot form regular patterns with five-fold symmetry that will fill space solidly and extend indefinitely in three dimensions. It is like trying to pave a floor with five-sided tiles'. This is illustrated in figure 72 and shows how this type of arrangement would explain the presence of short-range order and the absence of long-range order.

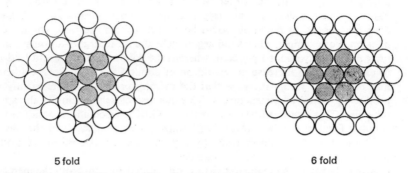

5 fold   6 fold

Figure 72. The difference between five-fold symmetry and six-fold symmetry, after Bernal.

In a later study Bernal emphasized the point that the attractive forces between molecules play little part in influencing the packing which occurs in the condensed state: the packing is almost entirely determined by the repulsive forces, that is the molecules behave very much like hard spheres. As Bernal remarks the key word (originally used in a different sense) is 'the one used by Humpty Dumpty in *Alice Through the Looking Glass*, impenetrability'. We are familiar with this in the solid state where, for practically all crystalline structures which are not dominated by covalent or other directed bondings, the arrangement is rather like that arising from packing the molecules as close as their repulsive forces will allow. For simple atoms, indeed, the crystal structure is like that of close-packed spheres. Bernal therefore considered the random packing, not of deformable plasticine spheres as in the earlier work, but the packing of a large number of hard steel balls. After introducing black paint and letting it harden the aggregate could be broken open and studied. In this way the number of contacts and near contacts could be determined. The model gives a co-ordination number varying between four and eleven, and Bernal considers this variation in contact number from molecule to molecule, to be the most significant feature of the irregular liquid structure. The packing very adequately reproduces the radial distribution function for liquid argon derived from neutron diffraction experiments (see figure 73), though this is probably not a very critical test.

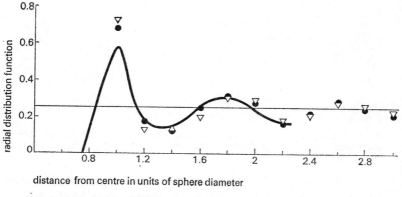

distance from centre in units of sphere diameter

——— from neutron diffraction
▽ ● calculated from random packing model

Figure 73. Radial distribution function for liquid argon plotted against distance from centre in units of sphere diameter. Results from neutron diffraction. O ▽ calculated from random packing model (● Bernal, ▽ Scott).

Further, the model does not allow vacancies between neighbours to be large enough to admit a third molecule.

The Bernal model aims to provide essentially an instantaneous snapshot (in say 10^{-15} sec.) of the structure of a liquid. If we now add to this the reality of thermal motion we see that the molecules are continuously changing the number of nearest neighbours, continuously shifting and shuffling around, and occasionally are able to allow one molecule to squeeze its way between neighbours without leaving a vacant hole. This aspect of the model may be open to criticism. For example, with a solid, because it is rigid, vacant sites or holes can exist in the lattice. The concentration of holes increases as the temperature is raised and by the time the solid reaches its melting point (but is not yet melted) about one site in 10^4 or 10^5 is vacant. Once melting occurs there is a further increase in volume, but according to Bernal the free space is more uniformly spread out. The X-ray evidence quoted above suggests that this is not possible, even at temperatures just above the melting point; in any case the thermal expansion of liquids is rather large so that at somewhat higher temperatures there must be vacancies or holes. Indeed, as Bernal himself points out, it is this increase in holes and the consequent drop in average co-ordination number that marks the transition from the coherent liquid phase to the incoherent gas phase (see figure 74). Bernal suggests that this probably occurs when the co-ordination number has fallen to an average value of three or four. Since the co-ordination number in the solid state (or in

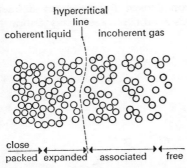

Figure 74. The transitions from the close-packed coherent liquid phase to the expanded, associated and finally free (incoherent gas) phase, as proposed by Bernal.

the liquid state near the freezing point) is about twelve, this means that the density of a gas at its critical point is between one quarter and one third of the density of the most highly condensed state. This is in fairly good agreement with observation for most *normal* liquids.

The Bernal approach is essentially that of a crystallographer; his model is a static one and concentrates on structural features. A far more analytical and difficult approach is that due to M. Born and H. S. Green. They carried out an extensive study of the mutual potential between molecules in the liquid state. They treat the molecules first in pairs, then in triplets and so on. The treatment is extremely complicated and the results are not easy to apply. In a sense their approach approximates to the theory of dense gases. A calculation is made of sets of co-ordinates of random collections of hard spheres. As the density of spheres is increased the behaviour passes from that of a gas to that of a dense gas and then to some other state. In this condition, by an iterative process, the co-ordinate positions can be repeatedly perturbed until a state of minimum energy is obtained. This corresponds over a critical density range, to the liquid state. The computation here is formidable even for only 100 spheres.

Yet another approach is that associated with the name of B. J. Alder; it is sometimes referred to as the method of molecular dynamics. It involves elaborate computations but the basic concept is simple. The molecules are treated as hard smooth perfectly elastic spheres, although as a later refinement an attractive square-well potential with a rigid core has been used. The molecules are placed in a box and started off with equal speeds in random directions and their subsequent motion determined by applying Newton's laws of motion. The boundary conditions chosen are such that a particle which passes out of one side of the box re-enters with the same velocity through the opposite side. In this way the number of molecules in the box and the total energy remain constant. The collisions within the system are analysed by a computer. It is found that after relatively few collisions the velocity distribution becomes very nearly Maxwellian. The total energy is determined by the initial velocities and of course remains constant; it determines the effective temperature of the system.

The behaviour depends on the velocity and on the packing of the particles. At one extreme the particles oscillate about an equilibrium position; this corresponds to the solid state. At the other extreme the particles move around throughout the box with relative freedom; this corresponds to the vapour regime. At some intermediate stage the particles vibrate about equilibrium positions but are also able to change places and slowly diffuse from one region to another. This corresponds to the liquid state.

We quote some typical results obtained by Alder and Wainwright (B. J. Alder and T. E. Wainwright, *Nuovo Cimento*, volume 9, 1958, supplement 1, p. 16 and, *Journal of Chemical Physics*, volume 31, 1959, p. 459), for 32 hard smooth particles in a box. They were able to map out a view of the calculated particle trajectories as seen from one face of the box. If v is the volume occupied by the particles and v_0 the volume they would occupy in a close-packed situation, they observed a very interesting instability for a ratio $v/v_0 = 1 \cdot 525$. For one group of collisions the particle trajectories appear as shown in figure 75(*a*) and calculation shows that the pressure on the walls is small; clearly

(a)   (b)

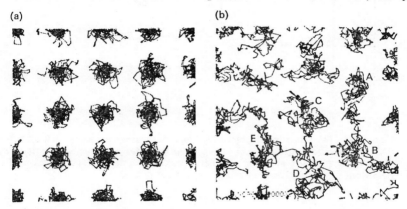

Figure 75.   End face view of box showing calculated particle trajectories for 32 hard elastic spheres (zero attractive forces) for a packing $V/V_0 = 1 \cdot 525$. (*a*) Behaviour resembling a solid. (*b*) Behaviour resembling a liquid. The particles spend most of their time oscillating within a *cage* of other particles, but are also able to swop places. See particles A and B; C, D and E.

this represents the solid state. In other groups of collisions the particle trajectories appear as in figure 75(*b*) and calculations show that the pressure on the walls is large. This is interpreted as the liquid phase. It is interesting to note that in the 'liquid' state the particles oscillate about a mean position but occasionally swop places (see figure 75(*b*), particles A and B particles C, D

203   The liquid state

and E). The situation shown in these figures refers to only 32 particles in the box. Apparently the system is not sufficiently large for the solid and liquid phases to exist in equilibrium: the system is in just that density range where part of the time it is in the liquid phase and part of the time in the solid phase.

Although the results reproduced in figure 75 are very striking, some reserve is necessary since an assembly of particles which show no attraction for one another can hardly be considered as a very realistic model of a liquid or a solid. However, from the point of view of packing, the structure scarcely depends on intermolecular forces (see Bernal above).

If the molecules are less closely packed the results obtained are as shown in figure 76. Figure 76(a) shows the potential for particles with a square-well

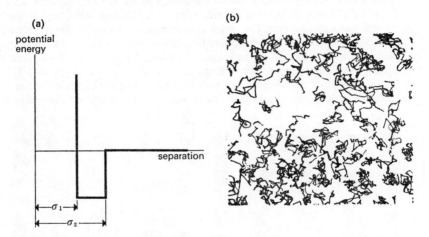

(a)

potential energy

separation

σ_1

σ_2

(b)

Figure 76. (a) Potential diagram of spheres possessing a square-well attractive field and a hard elastic core. The repulsive core has an effective diameter σ_1. (b) End face of box showing calculated trajectories of 108 such spheres. The behaviour of the particles resembles the vapour phase.

attractive field whilst figure 76(b) shows the particle trajectories for 108 such particles in a box; the behaviour evidently represents the liquid–vapour region. The Alder model brings out very clearly the dynamic features of the condensed state and the similarities and differences between the solid and liquid phases.

11·5   Ice and water

We introduce here a short digression on the relation between ice and water. Water, which is the basis of all our ideas of liquidity, is in a sense an atypical liquid. The forces between the molecules are very strongly directional; they arise from the interaction between electron-deficient hydrogen atoms in one molecule and electron-rich oxygen atoms in another. In the solid state this hydrogen bonding leads to a very open structure in which every water molecule is surrounded symmetrically (in the form of a regular tetrahedron) by four other water molecules. Consequently in ice the co-ordination number is four. As the temperature is raised the co-ordination number remains constant but the mean distance between the molecules increases. There is thus a small

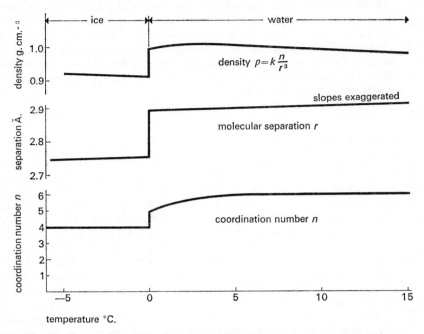

Figure 77.   The melting of ice. Schematic diagram showing (*a*) the change in co-ordination number, (*b*) the change in molecular separation, (*c*) the corresponding change in density as ice is heated from $-5°C$. to $+15°C$.

decrease in density with increasing temperature as with all 'normal' solids. At the melting point, however, a radical change occurs. The thermal energy is sufficient to destroy the regular bonding and the co-ordination number actually *increases*. At 0°C. the average co-ordination number appears to be between 4·5 and 5; with rising temperature this probably increases but soon reaches a constant value. This increase on its own must lead to a corresponding *rise* in density. However, an increase in temperature also leads to an overall increase in the mean distance between the molecules (thermal expansion). This expansion, on its own, would produce a *decrease* in density with increasing temperature. When combined with the change in co-ordination number it results in a 10 per cent jump in density on melting, a maximum density at 4°C., followed by a steady decrease in density at temperatures above about 10°C. The behaviour is shown schematically in figure 77. The details are by no means definite but the general pattern is reasonably valid. Another way of looking at this is in terms of the 'mixture-models' of water (e.g. H. S. Frank models). These stress the idea that water may not simply be a random or an ordered arrangement of individual molecules with uniform characteristics and bonding. On the contrary water may consist of two or more different molecular species, each species being structurally organized in its own distinctive way, with its own degree of co-ordination. The effect of temperature on this model is to increase the proportion of one species at the expense of the other. Whichever approach is adopted, the main point brought out by this discussion is the immense importance of the co-ordination number in determining the detailed properties of liquids.

11·6 General approach

In this chapter we shall treat the liquid as a modified solid, particularly when we are concerned with energies. When we consider flow we shall treat the liquid essentially as a solid, the molecules of which have a fair degree of mobility because of the presence of holes or free volume. This is the Eyring theory of viscous flow in a liquid; it differs from the Bernal theory in concentrating on thermal fluctuations rather than on the structural nature of the liquid state. It is much closer to the Alder model of the liquid state.

Before leaving this discussion we may make the following point. In the gaseous state the molecules are, more or less, completely independent. In the solid state they interact but the interaction, at least as far as crystalline solids are concerned, is capable of analysis precisely because of the long-range order which obtains. Solids are indeed an example of highly ordered co-operative interaction. By contrast the behaviour of a liquid is the result of co-operative interaction within a system of random order, or, at best, low-range order. It is this feature which makes the theory of liquids so difficult.

11·7 Latent heat of fusion

The latent heat of fusion is between one tenth and one thirtieth of the latent heat of sublimation. The latter represents the complete disintegration of the solid into the gas. Thus the change from solid to liquid involves a relatively small energy increase compared with the change from liquid to gas. For all the elements the latent heat of fusion is very small; for most compounds the latent heat of fusion corresponds approximately to the change from free vibration in the solid state to free rotation in the liquid. There is no detailed theory that is satisfactory. For many liquids the specific heat is of the order 3–6 cal. mole^{-1} °C.$^{-1}$

11·8   Melting point

The most astonishing feature of fusion is that the melting point is so sharp. If melting were always accompanied by a volume increase we could understand (if not explain) the sharpness of the melting point in the following way. Melting represents essentially the 'loosening-up' of the solid structure. If local melting occurs molecules on the edge of melted regions are exposed to molecules which are already a little further apart than usual. It would not be unreasonable to see in this the basis of a runaway situation where a little bit of melting soon engulfs the whole solid. If a pressure were applied to the solid to prevent a volume increase we should, in fact, find that the melting point is increased (Clausius–Clapeyron equation). Thus melting could cover a whole range of temperatures depending on the extent to which the volume increase is inhibited; only if the solid were completely free to expand would melting occur at a single sharply defined temperature. This 'explanation' would, of course, be a little embarrassing if we attempted to apply it to the melting of ice since here melting is accompanied by a decrease in volume.

As we saw earlier it is not possible to explain melting in terms of our potential energy curve, since this can only describe basically the solid state or the gaseous state. However, Lindemann (1910) suggested that one could obtain a good idea of the melting point from the P.E. curve by assuming that melting occurs when the amplitude of vibration exceeds a critical fraction of the atomic spacing. At this stage numerous collisions occur between neighbours and the crystal structure is destroyed.

Let x_0 be the vibration-amplitude when melting occurs. If f is the force constant between one atom (or molecule) and its neighbour and the motion is simple harmonic the mean total energy during vibration will be

$$u_v = \tfrac{1}{2} f x_0^2. \tag{11.9}$$

Since melting points are generally high we may assume the classical value for the internal energy, namely kT per atom or molecule for each principal direction of vibration. Thus at the melting point we write

$$u_v = kT_m = \tfrac{1}{2} f x_0^2. \tag{11.10}$$

But as we saw in an earlier chapter, for a simple cubic structure, $f = Ea$ where a is the atomic (or molecular) spacing and E is Young's modulus of the solid.

Hence $kT_m = \tfrac{1}{2} E a x_0^2.$

We assume that melting occurs when x_0 is some fraction of the atomic spacing a. Let x_0 be βa where $0 < \beta < 1$.

$$T_m = \frac{Ea^3\beta^2}{2k} = \frac{Ea^3 N\beta^2}{2R} = \frac{EV\beta^2}{2R}$$

where V = molecular volume, assuming a cubic structure.

$$T_m = \frac{E}{2\rho} \times \frac{M}{R}\,\beta^2, \tag{11.11}$$

where ρ = density of the solid.

Some typical results are given in Table 25 where it is assumed that β^2 has the value of $\frac{1}{50}$, i.e., $x_0 \simeq \dfrac{a}{7}$.

Table 25   *Melting point T_m (°K.) of some solids*

Solid	E dyne cm.$^{-2}$	ρ g. cm.$^{-2}$	M g.	T_m °K. Calculated	Observed
Lead	$1 \cdot 6 \times 10^{11}$	$11 \cdot 3$	207	400	600
Silver	$8 \cdot 3 \times 10^{11}$	$10 \cdot 5$	108	1100	1270
Iron	$21 \cdot 2 \times 10^{11}$	$7 \cdot 9$	56	1800	1800
Tungsten	$36 \cdot 0 \times 10^{11}$	$19 \cdot 3$	184	4200	3650
Sodium chloride	$4 \cdot 0 \times 10^{11}$	$2 \cdot 16$	57	1200	1070
Quartz	$7 \cdot 0 \ \times 10^{11}$	$2 \cdot 6$	60	1900	2000

This model has assumed simple cubic packing (so that $Na^3 = V$) and β a constant fractional amplitude of vibration at melting. Considering the crudity of the model and the range of materials to which it has been applied the agreement is surprisingly good.

11·9   Vapour pressure

11·9·1   *Molecular interpretation of vapour pressure*

Suppose we place a quantity of liquid in a sealed container and evacuate the air above it. Some of the liquid will evaporate since a certain fraction of the molecules will have enough excess thermal energy to escape from the attraction of neighbours. Some of these will return and condense. Equilibrium will be reached when the rate of evaporation equals the rate of condensation. The pressure exerted by the vapour under these conditions is known as the saturation vapour pressure or more usually the vapour pressure. It increases as the temperature is raised. The equilibrium vapour pressure will be almost identical in the presence of air or another gas (see however the section below on the effect of external pressure), since the vapour and the alien gas exert their partial pressures independently.

If the pressure above the liquid is set at some specified level, e.g. 1 atmosphere, the vapour pressure may be increased by raising the temperature of the liquid, until it is equal to the external pressure. At this point the liquid can evaporate freely: bubbles of vapour form copiously within the bulk of the liquid and the liquid is said to be boiling. The boiling point is thus the temperature at which the vapour pressure equals the external pressure – usually one atmosphere. If the external pressure is reduced the boiling point is lowered. It is a matter of experience that the boiling point on the absolute scale, at atmospheric pressure, is approximately two-thirds of the critical temperature, but there is no simple explanation for this.

The molecular interpretation of vapour pressure is simple. The molecules in the liquid have a wide range of thermal energies. The only molecules which can escape are those whose thermal energy is sufficient to overcome the attraction of neighbouring molecules. If the latent heat of vaporization per gram-molecule is L, the energy for a molecule to escape is

$$\varepsilon = \frac{L}{N_0}. \tag{11.12}$$

It follows that the average potential energy of the molecules in the vapour phase is greater than the mean value in the liquid phase by an amount ε. Consequently if n_V and n_L represent the number of molecules per unit volume in the vapour and liquid phases respectively and if the Boltzmann distribution holds for the liquid as well as for the vapour, we have

$$\frac{n_V}{n_L} = \exp\left(-\frac{\varepsilon}{kT}\right) = \exp\left(-\frac{L}{N_0 kT}\right) = \exp\left(-\frac{L}{RT}\right). \tag{11.13}$$

As the temperature is raised the ratio n_V/n_L increases but n_L scarcely changes with temperature. Since the vapour pressure p is directly proportional to the number of molecules per unit volume n_V in the vapour phase we may rewrite equation (11.13) as

$$p = A \exp\left(-\frac{L}{RT}\right), \tag{11.13a}$$

where A is a suitable constant. This relation is almost exactly correct.

We may approach the vapour pressure relation in another way. Suppose we consider the liquid to resemble a very dense gas containing n_L molecules per c.c. The number per c.c. with velocities, normal to the surface, between u and $u + du$ is given by equation (5.41) in Chapter 5. The number striking each sq. cm. of surface per sec. is then

$$u\,dn = n_L \left(\frac{m}{2\pi kT}\right)^{\frac{1}{2}} \int_0^\infty \exp\left(-\frac{mu^2}{2kT}\right) u\,du.$$

If we now specify that of these molecules only those with energy greater than ε (where $\varepsilon = L/N_0$) can escape, the number per sec. crossing each sq. cm. of surface into the vapour phase is the same integral as above but the limits are from $mu^2/2 = \varepsilon$ to ∞. This gives us the number escaping per sq. cm. per sec.,

$$n_L \left(\frac{kT}{2\pi m}\right)^{\frac{1}{2}} \exp\left(-\frac{\varepsilon}{kT}\right).$$

But if n_V represents the number of molecules per c.c. in the vapour phase, the number striking each sq. cm. of surface per second will be (see Chapter 5, equation (5.41))

$$n_V \left(\frac{kT}{2\pi m}\right)^{\frac{1}{2}}.$$

If all those molecules striking the liquid surface are adsorbed (this is the most dubious part of the treatment) then equilibrium occurs when the number escaping from the liquid exactly equals the number arriving from the vapour phase, that is

$$n_L \exp\left(-\frac{\varepsilon}{kT}\right) = n_V.$$

This is identical with equation (11.13).

11·9·2 *Effect of temperature on vapour pressure*

We use equation (11.13) to calculate the effect of temperature on vapour pressure. Since

$$p = A \exp\left(-\frac{L}{RT}\right),$$

we have by differentiating

$$\frac{dp}{dT} = A\frac{L}{RT^2}\exp\left(-\frac{L}{RT}\right) = \frac{L}{RT^2}\times p = \frac{L}{RT}\times\frac{p}{T}. \qquad (11.14)$$

If the vapour behaves like a perfect gas and V is the volume occupied by a gram-molecule of vapour at T we have $pV = RT$, so that

$$p = \frac{RT}{V}. \qquad (11.15)$$

Equation (11.14) becomes

$$\frac{dp}{dT} = \frac{L}{T}\times\frac{1}{V}. \qquad (11.16)$$

The correct relation, according to thermodynamics, is

$$\frac{dp}{dT} = \frac{L}{T}\frac{1}{(V-V_e)}, \qquad (11.17)$$

where V_e is the volume occupied by a gram-molecule of the liquid. This is known as the Clausius–Clapeyron equation. We may note that a plot of log p against T^{-1} should give a straight line of slope equal to L/R. This is indeed found to be so.

11·9·3 Effect of external pressure on vapour pressure

If we apply an external pressure, for example by compressing with an inert gas, how is the vapour pressure p of the liquid affected? The answer is surprising; p is increased. This may be explained in terms of figure 78 which

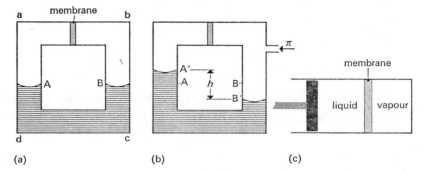

(a)   (b)   (c)

Figure 78. Effect of external pressure on the vapour pressure of a liquid. (a) Initial condition – the membrane is permeable only to the vapour. (b) Condition when external pressure π is applied. (c) Alternative model.

represents an idealized experiment. The top horizontal limb of the tube abcd contains a membrane permeable only to the vapour above the liquid AB. When the space above the liquid contains only vapour, A is level with B. If now an inert gas is pumped into the right-hand limb at a pressure π, since it cannot pass through the membrane, the level A rises to A′ and B falls to B′. If h is the difference in height, clearly

$$\pi = h\rho_L g. \tag{11.18}$$

Since the membrane is permeable to the vapour, the vapour over A′ must be in equilibrium with the vapour over B′. This implies that the vapour pressure at B′ is greater than over A′ by $h\rho_V g$. The increase in vapour pressure is thus

$$\Delta p = h\rho_V g = \pi \frac{\rho_V}{\rho_L}. \tag{11.19}$$

Another way of understanding this is to consider the arrangement in figure 78(c). The liquid is contained in a cylinder one end of which is closed by a movable piston. The bulk movement of the liquid is blocked by a membrane, which is impermeable to the liquid, but permeable to the vapour. The whole is maintained at a temperature T, so that heat is available for evaporation. If the piston is compressed with a pressure π until one gram-molecule of liquid has been forced to evaporate the liquid diminishes in volume by amount V_L; the piston does work πV_L on the liquid. Consequently the thermal energy necessary for evaporation is reduced from L to $L - \pi V_L$. The vapour pressure then becomes

$$p = A \exp\left[-\frac{(L - \pi V_L)}{RT}\right] = \left[A \exp\left(\frac{-L}{RT}\right)\right] \exp\left(\frac{\pi V_L}{RT}\right) \tag{11.20}$$

$$= p_0\left(1 + \frac{\pi V_L}{RT} + \ldots\right) = p_0 + \frac{p_0 \pi V_L}{RT}.$$

Hence

$$\Delta p = p - p_0 = \frac{p_0 \pi V_L}{RT}. \tag{11.21}$$

If the vapour behaves like a perfect gas $p_0 V_V = RT$ so that we may insert $p_0 = \frac{RT}{V_V}$ in equation (11.21). This gives

$$\Delta p = \pi \frac{V_L}{V_V} = \pi \frac{\rho_V}{\rho_L}. \tag{11.22}$$

Since at room temperature ρ_V/ρ_L is of the order of 10^{-3}, the increase in vapour pressure with external pressure is generally small. Descriptively one might say that the external pressure squeezes out the liquid molecules. It is not so easy to see why a negative pressure reduces the ease of escape of the liquid molecule. Nevertheless it is clear from the preceding experiment that a negative pressure will in practice reduce the vapour pressure.

11·10 Surface tension

11·10·1 *Surface energy*

It is common experience that liquids tend to draw up into drops; small drops form perfect spheres. Since a sphere is the geometric form which has the smallest surface area for a given volume we conclude that the surface of the liquid has a higher energy than the bulk; the liquid is therefore always attempting to attain its lowest energy state by reducing its surface area. This energy excess is called the free surface energy γ and is measured in erg cm.$^{-2}$

If a liquid has area A its free surface energy is γA. If we increase the surface by doing work on it the work done is

$$\frac{d(\gamma A)}{dA} = \gamma + A\frac{\partial \gamma}{\partial A}. \tag{11.23}$$

With a liquid, however much we stretch the surface the surface structure remains unchanged so that $\frac{\partial \gamma}{\partial A} = 0$. This is in contrast with a solid.

Hence

$$\text{Work done per unit area of extension} = \gamma. \qquad \textbf{(11.24)}$$

Recent experiments have shown that very thin films of water have the same surface energy as two surfaces separated by bulk water. The identity holds down to a thickness of only 20 Å. This implies that the molecular forces responsible for the surface energy (see below) are very short range.

The free surface energy is equivalent to a line tension acting in all directions parallel to the surface. Consider a liquid surface (as shown in figure 79) of

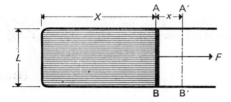

Figure 79. Sketch showing equivalence of surface energy per unit area and line tension per unit length.

width L and length X. Suppose at the barrier AB we apply a force F parallel to the surface and normal to AB so that we extend the length of the surface by an amount x. The area increase is Lx and the work done in increasing the surface area is $Lx\gamma$. This must be equal to the external work done Fx.

Hence $Fx = Lx\gamma$,

or $\qquad \frac{F}{L} = \gamma.$ \qquad \textbf{(11.25)}

Thus the surface energy is equivalent to a surface tension or line tension γ dyne cm.$^{-1}$

11·10·2 *Molecular interpretation*

The free surface energy of a liquid lends itself to a very simple molecular interpretation. Molecules in the bulk are subjected to attraction by surrounding molecules; the field is symmetrical and has no net effect. At the surface, however, the surface molecules are pulled in towards the bulk of the liquid, Apart from a few vapour molecules there is no attraction in the opposite direction. Consequently if we wish to increase the surface area we have to pull molecules up to the surface from the bulk against this one-sided attraction. This accounts for the surface energy.

213 The liquid state

Alternatively we may say that in moving molecules from the bulk towards the surface we are continuously breaking and reforming bonds. At the top, however, roughly half the bonds are not reformed. Since bond formation implies a decrease in energy the incomplete reformation of bonds at the surface means that the surface has a higher energy than the bulk. Practically the whole of this process is confined to the last few molecular layers and it is for this reason that extremely thin liquid films have the same surface energy as 'bulk' liquid (see above). Without going into further details it is evident that the surface energy is of the order of one half the energy required to break all bonds, i.e., one half the latent heat of vaporization of a molecular layer. This is essentially the same as the model we discussed earlier for solids and we emphasize again that the intermolecular forces responsible for the surface tension of a liquid are of the same nature as those responsible for the pressure defect of the corresponding gas.

If L is the latent heat of vaporization per mole and M the molecular weight and ρ the density, we have, as an approximate relation [cf. equation (7.9)],

$$\gamma = \frac{1}{2} \frac{L}{N_0} \left(\frac{N_0 \rho}{M} \right)^{\frac{2}{3}}. \tag{11.26}$$

For liquid argon this gives a value of 22 erg cm.$^{-2}$ for the surface tension compared with the observed value of 13; for liquid neon 6 compared with 5·5; for nitrogen 18 compared with 10·5; for oxygen 22 compared with 18; for benzene 180 compared with 40 and for mercury 1000 compared with 600 erg cm.$^{-2}$

Thermodynamics provides another quantity, the total surface energy h. When a surface is increased in area, apart from the surface energy term γ, there is also a 'latent heat' term. Heat must be provided to keep the temperature constant. Its magnitude is $-T \frac{\partial \gamma}{\partial T}$. Hence 'the total surface energy' is

$$h = \gamma - T \frac{\partial \gamma}{\partial T}. \tag{11.27}$$

Equation (11.26) does not distinguish between these two concepts; it is really a model for γ at 0°K. so that the poor agreement with the room temperature experimental values must not be taken too seriously. On the other hand the difference between h and γ is often significant. For example, for water $\gamma = 72$ and $h = 118$ erg cm.$^{-2}$ If fine droplets of water are allowed to coalesce so as to destroy their surface area the temperature rise is determined by h, not by γ. This provides a very ingenious method of determining areas of fine particles. They are first equilibriated with water vapour so that their surface is fully covered with a condensed film of water only a few angstroms thick. They are then immersed in water in a delicate calorimeter. If the heat given out is ΔQ erg, the surface area is given by $\Delta Q/118$ cm.^{2}

It is generally found that liquids with low surface tension readily wet most solids, giving a contact angle of zero; those with high surface tension often show a finite contact angle. In molecular terms we may say that if the cohesion between the molecules of the liquid is greater than the adhesion between the liquid and solid, the liquid will not 'want' to wet the solid, consequently it will show a finite contact angle.

Contact equilibrium is shown in figure 80. The free surface of the liquid has

(a)   (b)   (c)

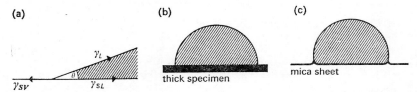

Figure 80. (*a*) Contact angle equilibrium. (*b*) Contact on a thick specimen. (*c*) Contact on a thin mica sheet where the vertical component of the tension distorts the shape of the sheet.

a surface energy γ_L; the solid liquid interface a surface energy γ_{SL} and the exposed portion of the solid adjacent to the liquid where vapour has been adsorbed a surface energy γ_{SV}. If we consider a virtual movement of the contact region so that an additional sq. cm. of solid is wetted we have

surface energy increase at solid–liquid interface $= \gamma_{SL}$,

surface energy decrease at solid–vapour interface $= \gamma_{SV}$,

surface energy increase of water surface $= \gamma_L \cos \theta$.

By the principle of virtual work

$$\gamma_{SV} = \gamma_{SL} + \gamma_L \cos \theta. \tag{11.28}$$

This equation, which was first derived by T. Young, without proof, in 1805 and by A. Dupré in 1869, is exactly what one would obtain if we took the horizontal components of the surface tension forces. This at once raises the question: what has happened to the vertical component $\gamma_L \sin \theta$? The answer is simple. The force is very small and its effect on a solid is negligible. If, however, the solid consists of a very thin flexible sheet, e.g., of mica, the vertical forces may be sufficient to distort the surface visibly. One is then aware of the reality of the vertical component.

Equation (11.28) has been studied experimentally by Miss S. Kay, using the bifurcation of mica as a means of determining directly the surface energies of the solid. In the first experiment she determined γ for a mica surface exposed to water vapour; this gives γ_{SV}. In the second experiment she determined γ

when the surfaces were immersed in water; this gives γ_{SL}. Since water completely wets mica ($\theta = 0$) we should expect to find, from equation (11.28) that

$$\gamma_L = \gamma_{SV} - \gamma_{SL}. \qquad (11.29)$$

Results obtained with water are given in Table 26. The table includes results for hexane which also wets mica.

Table 26 *Surface energy of mica in presence of vapour and liquid*

Fluid	Surface energies, erg cm.$^{-2}$			
	γ_{SV}	γ_{SL}	$\gamma_{SV} - \gamma_{SL}$	γ_L
Water	183	107	76	73
Hexane	271	255	16	18

The agreement is very satisfactory.

11·10·4 *Pressure difference across a liquid–vapour interface*

We shall now discuss a few of the general properties of capillarity which are shown by liquids. Before doing so we derive a simple relation for the pressure difference across a curved liquid interface.

Let ABCD be a portion of a surface marking the boundary between air below and liquid above (see figure 81). Over a small region a curved surface

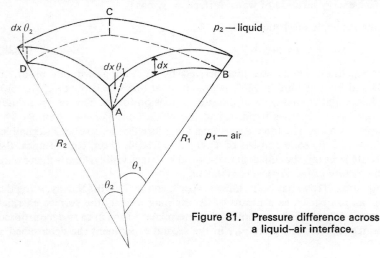

Figure 81. Pressure difference across a liquid–air interface.

can always be described in terms of two radii of curvature in mutually ortho-gonal planes. These are known as the principal radii of curvature R_1 and R_2. Suppose we assume that the pressure on the air side is p_1 and on the liquid side is p_2. We now apply the principle of virtual work displacing the surface parallel to itself through a distance dx. Then AB increases in length by $dx\,\theta_1 = dx\,\dfrac{\text{AB}}{R_1}$; similarly AD increases in length by $dx\,\theta_2 = dx\,\dfrac{\text{AD}}{R_2}$. The increase in surface area is then

$$\Delta\alpha = \left(\text{AB} + \text{AB}\frac{dx}{R_1}\right)\left(\text{BC} + \text{BC}\frac{dx}{R_2}\right) - \text{AB} \times \text{BC},$$

$$\Delta\alpha = \text{AB} \times \text{BC} \times dx \times \left(\frac{1}{R_1} + \frac{1}{R_2}\right). \tag{11.30}$$

This involves an increase in surface energy of $\gamma\Delta\alpha$. The work done by the pressure difference $p_1 - p_2$ is

$$(p_1 - p_2)\,\text{AB} \times \text{AD} \times dx = \gamma\Delta\alpha.$$

$$\therefore\, p_1 - p_2 = \gamma\left(\frac{1}{R_1} + \frac{1}{R_2}\right). \tag{11.31}$$

The pressure on the concave side is greater than on the convex side of a liquid interface, by an amount depending on the surface tension and on the curva-ture. This relation becomes clearer in the following two very simple cases. For a spherical bubble of radius R within a liquid the surface area is $4\pi R^2$. If a small expansion occurs increasing R by dR the increase in area of surface is $8\pi R\,dR$. The work that must be done by the pressure in the bubble is $4\pi R^2 p\,dR$. Equating this to the increase in surface energy we have

$$4\pi R^2 p\,dR = 8\pi R\,dR\gamma,$$

$$p = \frac{2\gamma}{R}. \tag{11.32}$$

This is exactly the result of equation (11.31) if $R_1 = R_2$. The second case is for a soap bubble. Here two surfaces are involved and the pressure inside the bubble is

$$p = \frac{4\gamma_s}{R}, \tag{11.33}$$

where γ_s is the surface tension of the soap solution.

11·10·5   Capillary rise

Suppose a uniform tube of small radius R is held vertically and lowered into a liquid of density ρ_L which wets it ($\theta = 0$). The liquid rises to a height h above the common level. This is sometimes explained in terms of the surface forces around the periphery of the meniscus. The meniscus has peripheral length $2\pi R$ and for zero contact angle the upward force is $2\pi R\gamma$. This is balanced by the downward weight of the liquid column $h\pi R^2 \rho_L g$ [figure 82(a)]. Thus

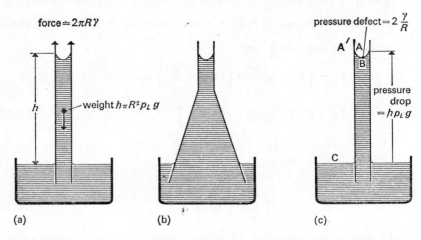

Figure 82. (a) Simple calculation of capillary rise in terms of surface tension forces. (b) Inapplicability of method a to a conical capillary. (c) Calculation of capillary rise in terms of pressure defect.

$$h\rho_L g = \frac{2\gamma}{R}. \tag{11.34}$$

Although equation (11.34) is correct, this is largely fortuitous. For example if the capillary was conical in shape [figure 82(b)] the capillary rise would still be given by equation (11.34) where R is the capillary radius at the point where the meniscus formed. Clearly $2\pi R\gamma$ could be considerably less than the weight of the column of liquid.

The proper method is to treat the problem in terms of the pressure difference across the meniscus. The meniscus for $\theta = 0$ is a hemisphere of radius $R_1 = R_2 = R$. According to equation (11.31) the pressure at B will be less than at A by the amount

$$\gamma\left(\frac{1}{R_1} + \frac{1}{R_2}\right) = \frac{2\gamma}{R}. \tag{11.35}$$

But the pressure at A is the same as that at A' (it is atmospheric) and is the same as that at C except for a minute air pressure drop A'C. Thus the pressure in the liquid at B is less than the pressure in the liquid at C. Consequently the liquid is drawn up the tube to a height h where

$$h\rho_L g = \frac{2\gamma}{R}.$$

We should obtain the same result for a conical capillary in agreement with experimental observation.

This treatment also explains why the meniscus in a narrow capillary can be considered to be spherical in shape. If the capillary rise is large compared with the diameter of the tube, the height of rise is practically constant over the whole width of the capillary, i.e., the pressure defect $h\rho_L g$ is constant. This means [see equation (11.35)] that the curvature of the meniscus must be constant. For a circular capillary the two principal radii of curvature must, by symmetry, be equal. Consequently the meniscus will be a portion of a sphere of almost constant radius, the value of which depends on the contact angle. For zero contact angle the shape will be hemispherical, the radius of curvature being equal to that of the tube R. If the contact angle is θ, it will be equal to $R/\cos\theta$ [see figure 83(a)] and the pressure defect giving rise to the capillary ascent will be equal to $\dfrac{2\gamma\cos\theta}{R}$. As the capillary gets wider the height of the

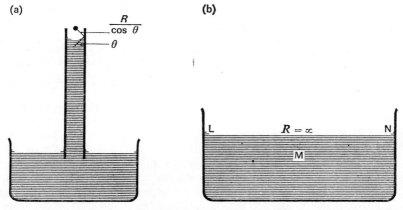

(a)   (b)

Figure 83. (a) Meniscus shape for large capillary rise must be of almost constant curvature. (b) For small capillary rise the radius of curvature becomes infinite at the centre.

liquid column becomes small compared with the capillary diameter [see figure 83(b)]. There is a very large difference between the pressure defect at L and at M. Indeed at M where the height of rise is negligible the curvature is almost zero ($R \to \infty$) whilst it is finite at L and N. The detailed shape of the meniscus can be solved theoretically by equating the pressure defect $\gamma\left(\dfrac{1}{R_1}+\dfrac{1}{R_2}\right)$ at any given point to the value of $h\rho_L g$ at that point.

219   The liquid state

Vapour pressure over a liquid meniscus

We see from figure 82(c) that the vapour pressure over the liquid at A is less than over the liquid at C by $h\rho_v g$ where ρ_v is the mean density of the vapour. We may write:

vapour pressure defect

$$= h\rho_v g = h\rho_L \times \frac{\rho_v}{\rho_L} \times g = \frac{2\gamma}{R} \times \frac{\rho_v}{\rho_L} . \tag{11.36}$$

This result is part of a more general statement: the vapour pressure over a concave meniscus surface is less than that over a flat liquid surface by the quantity given in equation (11.36). Similarly the vapour pressure over a convex meniscus is greater than over a flat surface by a similar amount. We may see this in a different way by considering the behaviour of a spherical liquid drop of radius R. The liquid within the drop is subjected to a surface tension pressure of amount $p = 2\gamma/R$. As we saw in a previous section this increases the vapour pressure by the amount

$$p \frac{\rho_v}{\rho_L} = \frac{2\gamma}{R} \times \frac{\rho_v}{\rho_L} . \tag{11.37}$$

11·11 Nucleation in condensation: the Wilson cloud chamber

Consider the equilibrium between a vapour and its liquid and the problem of condensation. If there are already present a few droplets of radius of curvature r, the droplets exert a vapour pressure of $\frac{2\gamma}{r} \times \frac{\rho_v}{\rho_L}$. This can be very large. For example with water vapour at 20°C., ρ_v/ρ_L is about 2×10^{-5}. But if r is only 5 Å. the factor $2\gamma/r$ has a value of about 3×10^9 so that the droplets are attempting to evaporate with a vapour pressure of about 6×10^4 dyne cm.$^{-2}$ Since the saturated vapour pressure of water at 20°C. is about 2×10^4 dyne cm.$^{-2}$, the droplets will evaporate; they cannot grow by further condensation. There are only two ways by which condensation can be made to occur. One is to provide artificial nuclei with relatively large radii of curvature: this is, in practice, generally achieved by dust particles. The second is to produce supersaturation of the vapour, usually by sudden cooling. This is the basis of the Wilson cloud chamber.

Consider a vessel closed with a moveable piston in which the vapour is in equilibrium with the liquid. Charged particles are projected through the vapour. Because the electrostatic energy of a point charge is lowered by the condensation of a dielectric on it, some condensation may occur. If the piston is now suddenly raised there is a marked adiabatic temperature drop. Although there is some volume increase the main effect is that the existing vapour pressure is much higher than the saturated vapour pressure at the new reduced temperature. Condensation occurs readily and the path of the particles is revealed as a series of liquid droplets.

11·12 Superheating

A similar problem is involved in the heating of a liquid to produce boiling. If the initial bubbles have a radius of curvature r, the vapour pressure in the bubbles must exceed $2\gamma/r$ if the bubbles are to grow. For a bubble of radius $r = 5$ Å. in a liquid such as water, this pressure is of the order of 3000 atmospheres. Thus unless the vapour pressure reaches this value the bubble will collapse. To obtain vapour pressures of this order the liquid must be very strongly superheated. With water, temperatures of the order of several 100°C. are required. Experiments with water from which dissolved gases have been eliminated do, in fact, show that bubbles will not form until temperatures of this order are reached. Of course, once the bubble begins to form at this high vapour pressure, the excess pressure necessary to produce bubble growth begins to decrease as r increases, so that the bubble grows at a catastrophic rate – the liquid explodes.

Steady boiling at the true boiling temperature is achieved by providing suitable nuclei. These are often air pockets trapped at solid surfaces or gases dissolved in the liquid itself.

11·13 The energy for capillary rise

We raise here two points concerning capillary rise that are not often discussed in elementary textbooks. The first is, what is the source of the energy for capillary rise? Consider a tube of radius R. The liquid rises to a height h given by

$$h = \frac{2\gamma_L}{R\rho g} . \tag{11.38}$$

The potential energy ε_1 gained by the liquid is equivalent to the raising of a mass of liquid $\pi R^2 h\rho$ through a height $\frac{h}{2}$. Hence

$$\varepsilon_1 = \tfrac{1}{2} \pi R^2 h^2 \rho g. \tag{11.39}$$

This energy comes from the wetting of the walls of the tube by the liquid˙ Descriptively we may see this in the following way. The surface energy of a solid arises from the asymmetric forces at the free surface. If we cover the surface with another material, e.g. a liquid, we reduce this asymmetry and hence reduce the surface energy. The energy liberated in this process is available for pulling the liquid up the tube. We may make this quantitative as follows.

Consider a liquid with contact angle $\theta = 0$. The equilibrium condition, as we saw earlier, is

$$\gamma_{SV} = \gamma_L + \gamma_{SL}. \tag{11.40}$$

If the liquid advances along the tube so as to cover 1 sq. cm. of the surface we destroy one sq. cm. of γ_{SV} and gain one sq. cm. of γ_{SL}. There is no change in the area of the liquid meniscus. The energy given up by the system is thus

$$\gamma_{SV} - \gamma_{SL}. \tag{11.41}$$

From equation (11.40) we see that this is equal to γ_L. Thus each sq. cm. of capillary surface wetted releases energy γ_L. The energy released for a rise of h is then

$$\varepsilon_2 = \gamma_L \, 2\pi Rh. \tag{11.42}$$

Substituting from equation (11.38) for γ_L we have

$$\text{loss of surface energy } \varepsilon_2 = \pi R^2 h^2 \rho g. \tag{11.43}$$

This is exactly twice the gain in potential energy given by equation (11.39). This implies that if the liquid were non-viscous it would rise to a height $2h$ and oscillate between 0 and $2h$ with its mean position at h. In practice viscosity consumes the excess energy very quickly.

The behaviour is like that of a vertical spiral spring on to the end of which a weight is suddenly attached. The weight falls instantaneously to double its equilibrium depth and oscillates between this and zero until the kinetic energy is consumed by friction and static equilibrium is achieved.

The second query is the following. Some solids are known to have much higher surface energies than others. For example if they are first thoroughly cleaned and then immersed in a liquid the amount of heat released may vary very widely; nevertheless if a liquid wets these surfaces ($\theta = 0$), the capillary forces do not depend on the material of which the capillary is made; they depend only on the surface energy of the liquid. Why is this? The answer is implicit in equation (11.40). The adsorbed vapour on the solid will always adjust its thickness so that the quantity,

$$\gamma_{SV} - \gamma_{SL},$$

is equal to γ_L (for $\theta = 0$). The energy gain is thus dependent only on γ_L. We may describe this differently. The vapour adsorbed on the solid presents a surface to the advancing liquid which closely resembles the liquid itself. On a high-energy surface the adsorbed film will tend to be thicker than on a low-energy surface. In both cases (for $\theta = 0$) the liquid is virtually wetting itself so that only γ_L is involved in the process. In this connection we may note that the range of forces is small and that the adsorbed film needs to be only 10 to 20 Å. thick in order to mask almost completely the influence of the underlying substrate.

Chapter 12
Liquids: their flow properties

In this chapter we shall discuss the flow properties of liquids, and derive or state some of the more standard equations of flow. However, our main attention will be directed to a molecular model of viscous flow; this follows closely the theory proposed by Eyring.

12·1 Flow in ideal liquids: Bernoulli's equation

As with gases we can assume the existence of ideal liquids in which internal forces play a trivial part, so that they have negligible surface tension or viscosity. Their flow properties are determined solely by their density. Of course, no liquids can have zero internal forces, they would not be liquid if this were so, but they can often behave as ideal liquids in flow if the inertial forces dominate.

Consider the flow of such an idealized liquid and let us follow the path of any particle in it. If it moves in a continuous steady state we may draw a line such that the tangent at any point gives the direction of flow of the particle. Such lines are called streamlines. They are smooth continuous lines throughout the liquid and can never intersect – no fluid particles can flow across from one streamline to another. Let AB represent an imaginary tube in the liquid bounded by streamlines [figure 84(a)]. At A the liquid is at a height h_1 above

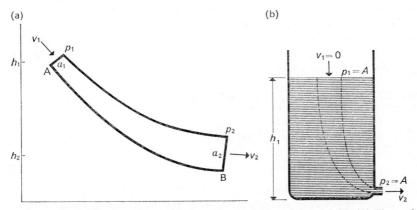

Figure 84. (a) Streamlines of flow of a liquid in a gravitational field. (b) Flow of liquid from an orifice at a depth h below the free level.

a reference level, the pressure acting at that point is p_1, the cross-sectional area of the tube is α_1, and the liquid flow velocity is v_1; at B the corresponding quantities are h_2, p_2, α_2, v_2. Consider the energy balance during a short time interval dt. The pressure p_1 at A drives a volume of liquid $\alpha_1 v_1 dt$ through the tube: the work done is

$$w_1 = p_1 \alpha_1 v_1 dt. \tag{12.1}$$

The mass of liquid involved is $\rho \alpha_1 v_1 dt$ so that the kinetic energy it brings with it is

$$w_2 = \tfrac{1}{2}(\rho \alpha_1 v_1 dt)v_1^2, \tag{12.2}$$

and its potential energy is

$$w_3 = (\rho \alpha_1 v_1 dt)gh_1. \tag{12.3}$$

Since the motion is assumed steady and the liquid is ideal the same amount of energy leaves the tube at B. Hence

$$(w_1 + w_2 + w_3)_A = (w_1 + w_2 + w_3)_B. \tag{12.4}$$

Because the same mass of liquid leaves at B as enters at A,

$$\alpha_1 v_1 dt = \alpha_2 v_2 dt. \tag{12.5}$$

Equation (12.4) becomes

$$\frac{p_1}{\rho} + \tfrac{1}{2}v_1^2 + h_1 g = \frac{p_2}{\rho} + \tfrac{1}{2}v_2^2 + h_2 g. \tag{12.6}$$

This is Bernoulli's equation of flow for an ideal liquid: the three terms on each side of the equation correspond respectively to the pressure energy to move the liquid, the kinetic energy and the potential energy.

A simple example is illustrated in figure 84(b). A container is filled with liquid to a height h_1. At the bottom of the container is a small hole. What is the velocity v_2 of the outflowing liquid? If we construct a tube of flow as shown the conditions at the top of the tube are $p_1 =$ atmospheric $= A$, height $= h_1$, velocity $v_1 \simeq 0$. At the bottom of the tube $p_2 = A$, height $= 0$, velocity $= v_2$. Starting with equation (12.6),

$$\frac{p_1}{\rho} + \tfrac{1}{2}v_1^2 + h_1 g = \frac{p_2}{\rho} + \tfrac{1}{2}v_2^2 + h_2 g,$$

$$\frac{A}{\rho} + 0 + h_1 g = \frac{A}{\rho} + \tfrac{1}{2}v_2^2 + 0,$$

$$v_2^2 = 2gh_1. \tag{12.7}$$

The flow of an ideal liquid does not involve any particular molecular model; density is the only property concerned, and we shall not discuss it further. We may, however, remark that in general an increase in flow velocity is accompanied by a drop in pressure. This accounts for the suction experienced by a ship approaching a quay – the water flows faster at the regions of nearest approach. It also accounts for the downward curvature of the flight of a tennis ball that has been given top spin. Again, because the top surface of an aerofoil is longer, the air velocity in flight is higher and the pressure lower than over the bottom surface; this produces aerodynamic lift.

12·2 Flow in real liquids, viscosity

12·2·1 *Steady and unsteady flow, turbulence*

Real liquids experience a viscous resistance to flow. We have already defined viscosity in discussing the behaviour of gases but we may recapitulate here. If the velocity gradient between two neighbouring planes is du/dz the force per sq. cm. to overcome viscous resistance is

$$f = \eta \frac{du}{dz},$$ (12.8)

where η is defined as the viscosity. Fluids which obey this relationship are known as Newtonian fluids. The dimensions of η are

$$[\eta] = ML^{-1}T^{-1}.$$ (12.9)

If the flow is steady and stable the work done is expended solely in overcoming the viscosity, and the work appears as heat. If, however, the velocity of flow is too high or there are other unfavourable features, a series of vortices may develop in the liquid. Some of the work is expended in providing kinetic energy for the vortices. Vortices often assemble on a solid boundary and this is known as the boundary layer. The condition for turbulent motion was first established by O. Reynolds. If a liquid of density ρ, viscosity η flows along a channel of lateral dimension a cm., there is a critical velocity at which orderly streamlined flow changes to turbulent motion. By dimensional analysis it is easy to show that this occurs when

$$v_c = \frac{k\eta}{\rho a}.$$ (12.10)

The quantity k is called Reynold's number. It is dimensionless and is fairly constant over a wide range of conditions. Its value is of the order of 1000.

For $\dfrac{v\rho a}{\eta} < k$ the flow is streamlined.

For $\dfrac{v\rho a}{\eta} > k$ the flow is turbulent.

Turbulence is essentially a *condition* of instability and the force involved in flow is not necessarily a clear indication of whether turbulence has begun or not. The force is determined primarily by viscous or inertial factors. An interesting simple example of these two conditions is furnished by considering the steady movement of a solid sphere of radius a through a fluid. The resistance to motion can be derived by dimensional arguments. One solution for the resistive force is

$$F = Aa\eta v. \tag{12.11}$$

This is Stokes' law; analysis shows that A has the value 6π.

Another solution is

$$F = Bv^2\rho a^2. \tag{12.12}$$

This does not involve viscosity but kinetic energy, the force being determined by the momentum of the oncoming fluid.

12·2·2 Viscous flow through a tube

Consider a cylindrical tube of length l, radius r. Liquid enters one end at pressure p_1 and leaves the other end at pressure p_2. If the flow is uniform the streamlines are all parallel to the tube axis, there is no slip of liquid at the solid boundary and the velocity profile is parabolic. The rate of flow of liquid per sec. is

$$Q = \frac{\pi r^4}{8\eta}\left(\frac{p_2 - p_1}{l}\right). \tag{12.13}$$

Poiseuille first used this equation to determine the viscosity of blood in horses' arteries.

12·2·3 Rotation of a shaft in a bearing: hydrodynamic lubrication

Consider a shaft of radius a rotating with angular velocity ω in a bearing of radius $a+c$ [figure 85(a)]. If the shaft carries a light load W and the space between shaft and bearing is filled with oil, they will remain concentric. The velocity gradient across the film is $\omega a/c$. The viscous force f per sq. cm. is this quantity multiplied by η. If the length of the shaft is l, the resultant tangential force that must be exerted at the edge of the shaft to maintain steady rotation is

$$F = f \times 2\pi al = \frac{\omega a}{c}\eta\, 2\pi al = \frac{Na}{c}\eta al \times 4\pi^2, \tag{12.14}$$

where $N = \dfrac{\omega}{2\pi}$ is the number of revolutions per sec.

We define the coefficient of friction μ as the ratio of F to W. Then

$$\mu = \frac{Na}{c} \times \eta\, \frac{al}{W} \times 4\pi^2. \tag{12.15}$$

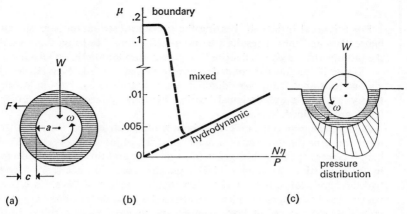

(a) (b) (c)

Figure 85. (a) Rotation of a shaft of radius a in a bearing of radius $a+c$ (Petroff). If the liquid has viscosity η and the angular velocity is ω the viscous force F that has to be overcome is proportional to $\eta\omega$. (b) If the normal load is W the coefficient of friction is proportional to $\dfrac{\eta\omega}{F}$ or to $\dfrac{\eta N}{P}$ where N is the number of revolutions per second and P the normal pressure. At high pressures or low speeds the hydro-dynamic film is penetrated and the friction rises. (c) In practice a convergent oil wedge must be formed so that sufficient pressure can be developed in the liquid to support the applied load; as a result the shaft is not quite concentric with the bearing (Reynolds).

Since the nominal pressure P on the bearing is $\dfrac{W}{2al}$ we may write

$$\mu = \frac{4\pi^2 N\eta}{P}\left(\frac{a}{2c}\right).$$ (12.16)

For a bearing and shaft of fixed dimensions the quantity $\dfrac{a}{2c}$ is a constant so that μ is linearly proportional to $N\eta/p$ [see figure 85(b)].

This relation was first derived by A. A. Petroff in 1883 and describes adequately the behaviour of a lightly loaded bearing. However, it is evident that for loads which are not negligible, the journal cannot run in the central position. If the oil film is to support an appreciable load the journal must rotate eccentrically relative to the bearing, so that the lubricant is squeezed through the converging gap between the surfaces. This convergence in a viscous liquid, produces a hydrodynamic pressure in the oil wedge as O. Reynolds showed in 1886 and under suitable conditions it is sufficient to support the applied load.*

* If the liquid were completely non-viscous the rotating journal would not drag any liquid with it. If we were to supply an external pressure and force the liquid through the gap, the convergence, as we saw in section 12.1, would actually produce a pressure *drop* in the convergent gap. It is primarily the viscous property of the liquid that is responsible for the development of a positive pressure in the liquid wedge.

For a journal in a 'half' bearing the pressure distribution is as shown in figure 85(c) and similar results are obtained for a 'full' bearing. It is seen that the (asymmetric) pressure distribution gives a resultant upthrust which would not occur if the journal rotated centrally in the bearing. Nevertheless the viscous resistance is not very different from that given by the simple Petroff theory, and for a well-designed bearing friction coefficients of the order of 0·001 are obtained.

A far more important difference between the Petroff and Reynolds treatment concerns the distance of nearest approach of journal to bearing. As the load increases the oil convergence must increase and this leads to a decrease in the separation. At some critical stage the separation necessary for ideal hydrodynamic lubrication may be less than the height of the surface roughnesses. The hydrodynamic film is penetrated and contact occurs between the metals themselves or, at best, between metals covered by a thin adsorbed lubricant layer only one or two molecules thick. This is referred to as boundary lubrication and involves higher friction ($\mu = 0\cdot1$) and possible wear. By contrast hydrodynamic lubrication gives a very low coefficient of friction ($\mu \sim 0\cdot001$) and, in principle, no wear at all. The region between the classical hydrodynamic and boundary regime is often referred to as the mixed regime. More recent work shows that in this region contact between asperities may be prevented by another mechanism. The very high pressures generated at the points of imminent contact can produce a prodigious local rise in viscosity. Consequently the lubricant cannot be extruded from precisely those areas where metallic contact might otherwise occur. In this way the range of successful lubrication is extended into a region of far more severe conditions. This type of lubrication, where high elastic stresses have a favourable effect on the viscosity of the lubricant, is known as elasto-hydrodynamic lubrication. When the elasto-hydrodynamic film itself is finally penetrated or destroyed we are back again at classical boundary lubrication where only molecular adsorbed layers remain.

12·2·4   *Simple ideas on the viscosity of liquids*

The viscosity of gases is explained in terms of molecular momentum-transfer from one plane to a neighbouring plane. The mechanical analogy is to consider two trains travelling parallel to one another at slightly different speeds. People are continuously jumping from one train to the other, so that those leaving the slower train, slow up the faster; whilst those jumping from the faster train speed up the slower. There is no net change in the numbers of people in any one train but there is clearly a momentum transfer tending to equalize the velocities of the two trains. Work must be done on the trains to maintain their difference in velocity. It is this process on a molecular scale which is responsible for the viscosity of gases.

In liquids the momentum transfer process also takes place but it is a trivial part of the viscous resistance. The train model may still be used but now the jumping passengers find themselves caught by their coat-tails. They must grab hold of the other train and pull very hard before they can detach themselves from their own train. This will have a marked effect in tending to equalize the train velocities. In molecular terms one may say that in a liquid the molecules are so close together that considerable energy must be expended in dragging one molecular layer over its neighbour. No rigorous theory exists but a simple molecular model may be given. We first give it descriptively and in the next section deal with it analytically.

Consider an instantaneous picture of the liquid. There is short-range order and we may represent a small part of two neighbouring molecular planes by an array similar to that of a solid (figure 86). To drag molecule A2 from its

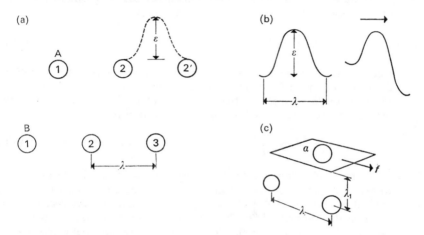

Figure 86. The Eyring molecular model of liquid viscosity (a) to transpose a molecule from position A2 to A2' intermolecular forces must be overcome. This involves a potential energy barrier ε, (b) if a shear stress is applied to the right the potential energy curve is distorted, transitions to the right are facilitated, to the left hindered, (c) a shear stress f acting on a molecule occupying a nominal area α exerts a shear force $f\alpha$.

position to the neighbouring equilibrium site A2' involves overcoming the attractive forces of B2. There is thus a potential hill to be climbed, and energy is needed for this. (Note that in falling from the summit X to site A2' the energy that is restored appears as heat.) Occasionally molecules will have enough thermal energy to do this but there is just as much chance of the molecule jumping to site A1 as to A2'. There is thus no net flow in any direction. If now we apply a shear stress to the right on the molecules in plane A mechanical work is done on the molecules. The thermal energy necessary to move the molecule to the right is reduced whilst that required to move it to the left is increased [figure 86(b)]. Thus thermally activated flow to the right

229 Liquids: their flow properties

is favoured. Another way of looking at the process is to say that before molecule A2 can move to site A2' it must make available a vacancy in the liquid to receive it. The energy to do this depends on the amount of opening up of the liquid that is necessary and on the latent heat of vaporization. Indeed the energy needed to make a vacancy available is roughly equal to $\frac{1}{2}$ to $\frac{1}{3}$ of the latent heat of vaporization of the vacancy.

12·2·5 *The Eyring theory of liquid viscosity*

We consider a small cluster of molecules as shown in figure 86. In order to transpose molecule A2 to position A2' against the attractive forces of neighbours we have to surmount a potential energy barrier ε. The rate at which this is possible as a result of thermal fluctuations is

$$C \exp\left(-\frac{\varepsilon}{kT}\right),$$ (12.17)

where C is a frequency term. Eyring suggests that this is the vibrational frequency of the molecule and is given approximately by kT/h, where h is Planck's constant. Hence the frequency with which a molecule can jump to a neighbouring site as a result of thermal activation is

$$\mathcal{N}_0 = \frac{kT}{h} \exp\frac{-\varepsilon}{kT}.$$ (12.18)

When no stress is applied to the liquid there is an equal rate of jumping to left and right so that no net flow occurs. Suppose now a shear stress f is applied [see figure 86(c)]. If the average area occupied by a molecule is α the force on the molecule is $f\alpha$. The only mechanical work we can perform is expended in carrying the molecule to the top of the potential barrier (on the other side of the barrier peak the molecule is assumed to give up all its energy as heat). The work done is

$$\frac{f\alpha\lambda}{2},$$ (12.19)

where λ is the distance between neighbouring molecules in the plane.

It follows that the thermal energy required for a jump to the right is reduced from ε to $\varepsilon - \frac{f\alpha\lambda}{2}$, i.e., the rate of jumping is increased. Similarly the thermal energy required for a jump to the left is increased to $\varepsilon + \frac{f\alpha\lambda}{2}$ so that the jumping rate is reduced. The net rate of jumping to the right is now,

$$\mathcal{N} = \frac{kT}{h}\left[\exp\left(-\frac{\varepsilon - \frac{f\alpha\lambda}{2}}{kT}\right) - \exp\left(-\frac{\varepsilon + \frac{f\alpha\lambda}{2}}{kT}\right)\right]$$ (12.20)

$$\mathcal{N} = \frac{kT}{h} \exp\left(-\frac{\varepsilon}{kT}\right) \left[\exp\left(\frac{f\alpha\lambda}{2kT}\right) - \exp\left(-\frac{f\alpha\lambda}{2kT}\right)\right]. \qquad \textbf{(12.21)}$$

For the exponentials in the square bracket we may take the first terms of the expansion. This gives

$$\mathcal{N} = \frac{kT}{h} \exp\left(-\frac{\varepsilon}{kT}\right) \left[2\frac{f\alpha\lambda}{2kT}\right] = \frac{f\alpha\lambda}{h} \exp\left(-\frac{\varepsilon}{kT}\right). \qquad \textbf{(12.22)}$$

We must now link $\mathcal{N}$ with the macroscopic flow process. The velocity gradient across the two molecular layers separated by a distance λ_1 is

$$\text{velocity gradient} = \frac{\text{velocity difference}}{\lambda_1}$$

$$= \frac{(\text{distance per jump}) \times (\text{number of jumps per sec.})}{\lambda_1}$$

$$= \frac{\lambda\mathcal{N}}{\lambda_1}. \qquad \textbf{(12.23)}$$

But by definition the viscosity η is given by

$$\eta = \frac{\text{force per unit area}}{\text{velocity gradient}} = f \times \frac{\lambda_1}{\lambda\mathcal{N}}. \qquad \textbf{(12.24)}$$

Inserting the value of $\mathcal{N}$ from equation **(12.22)** we obtain

$$\eta = \frac{h\lambda_1}{\alpha\lambda^2} \exp\left(\frac{\varepsilon}{kT}\right). \qquad \textbf{(12.25)}$$

Putting $\lambda_1 \simeq \lambda$ and treating $\alpha\lambda_1$ as the effective volume occupied by a molecule (v_m), equation **(12.25)** becomes

$$\boxed{\eta = \frac{h}{v_m} \exp\left(\frac{\varepsilon}{kT}\right)}$$

or $\qquad \eta = \frac{hN_0}{V_m} \exp\left(\frac{E}{RT}\right),$ $\qquad\qquad\qquad$ **(12.**

where N_0 is Avogadro's number, V_m the volume per mole of the liquid and E the molar activation energy for surmounting the energy barrier. This relation is in full accord with the very marked decrease in viscosity observed with increasing temperature.

The barrier energy E may also be regarded as the energy to create a hole in the liquid big enough to receive a molecule. This ought, therefore, to be comparable with the latent heat of vaporization (L); however, because there is already some free volume in the liquid, the work to open up a molecular hole is less than this. For a very large range of liquids

$$E \simeq (0\cdot3 \text{ to } 0\cdot4)L.$$

Consequently for a very wide range of liquids we may write

$$\eta = \frac{hN_0}{V_m} \exp \left(\frac{0.4L}{RT} \right) . \tag{12.27}$$

However for liquid metals it is found that E is very much less than $0.4\,L$, presumably because the metal ions are much smaller than the metal atoms.

12·2·6   *Effect of external pressure on viscosity*

To examine the effect of pressure on liquid viscosity it is simplest to regard the energy barrier E as the energy required to open up a hole for the transposed molecule. In the absence of applied pressure we may write equation (12.26) in the form

$$\eta_0 = D \exp \left(\frac{E}{RT} \right) . \tag{12.28}$$

If now an external pressure P is applied the work to form a hole is increased by PV_h where V_h is the volume of the hole. The thermal energy for activated flow is then increased from E to $E+PV_h$. The viscosity then becomes

$$\eta = D \exp \left(\frac{E+PV_h}{RT} \right)$$

$$= \eta_0 \exp \left(\frac{PV_h}{RT} \right) . \tag{12.29}$$

This type of relation is found to hold over a fairly wide range of pressures. If $\log_e \eta$ is plotted against P a straight line of slope $\dfrac{V_h}{RT}$ is obtained. It turns out that V_h is about 15 per cent V_m for simple liquids and about 5 per cent V_m for liquid metals. Attempts to explain away these discrepancies are not convincing. One must accept this as part of the inadequacy of the model. In broad outline it is surprisingly successful.

The effect of high pressures in increasing viscosity can be very significant in lubrication. It is probable that many sliding mechanisms operate successfully because local high pressures, which would normally be expected to squeeze out the lubricant and cause seizure, so increase the viscosity of the oil that it remains trapped between the surfaces. For example with heavily loaded gear teeth where elastic deformation in the contact zone may produce pressures of the order of 100,000 p.s.i., the viscosity of a typical lubricant may be increased a million-fold. In this way nature is much kinder to the engineer than he might have supposed. This type of elasto-hydrodynamic lubrication, (mentioned previously on page 228) is an area of great importance in current lubrication practice.

Although the model of liquid viscosity which we have discussed above is greatly simplified it brings out three important points. Liquid viscosity arises from intermolecular forces and the ability of thermal fluctuations aided by an applied shear stress to produce flow in the direction of shear. Consequently the viscosity decreases if the temperature is increased. Secondly the viscosity increases if the pressure on the liquid is increased. This is because the molecules are pushed closer together, or, in terms of the alternative way of describing the molecular jumps, more work must be done to open up a vacant site.

A third conclusion is that viscosity, being determined by intermolecular forces, is closely related to surface tension γ and both are related to the latent heat of vaporization. For many simple organic liquids there is a fair correlation between log η and γ/Z where Z is a factor allowing for the size of the molecule.

It may be mentioned that these models are most suitable for unassociated liquids with van der Waals bonding between the molecules. Highly polar molecules and long complicated molecules do not fit in so well.

The basic differences between the mechanisms responsible for viscosity in gases and in liquids are clearly brought out by the following table.

Table 26   *Viscosity of gases and liquids*

Fluid	*Effect of temperature T*	*Effect of pressure P*
Gases	η increases as $T^{\frac{1}{2}}$	None
Liquids	η decreases as $\log_e \eta = A + \dfrac{B}{T}$	η increases as $\log_e \eta = A + kP$

12·3   Rigidity of liquids

The molecules in a liquid jump about with a frequency of 10^{10} to 10^{12} per sec. If the rate of shear is sufficiently great there may not be time for the molecules to advance to a neighbouring site. In this case the liquid will not show viscous flow, but will show a finite elastic rigidity. For simple liquids the rates of shear for this to occur are enormous but for liquids with large bulky molecules it may occur at moderate rates of shear. Silicone putties are a good example of this. There is a similar behaviour in polymers. The flow of polymeric chains which occurs at slow rates of loading resembles the viscous flow of liquids; this may be impossible at high rates of strain. The material will then deform elastically. Silicone putty for example will bounce with a very high resilience, since there is not enough time for the molecules to flow. If the material is in the form of a rod and is pulled rapidly it will first stretch and then snap in a brittle fashion. On the other hand at very low rates of deformation, it will flow in a viscous manner.

12·4 Non-Newtonian flow

Many liquids encountered in daily life are non-Newtonian. This is particularly true of liquids with large bulky molecules or those which have some additional structure, e.g. colloids. A finite stress is often needed to break down the structure and initiate flow [see figure 87(a)]. Such materials are known as

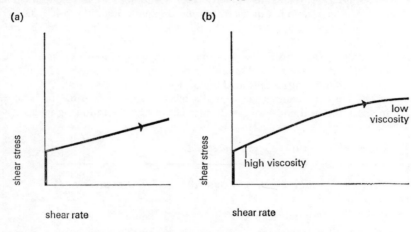

Figure 87. Schematic diagram of shear stress against shear-rate for (a) a Bingham fluid, (b) a Bingham fluid showing thixotropy.

Bingham fluids. In many cases, once flow starts the molecules tend to orient along the direction of shear. The interaction of the molecules is reduced and the viscosity diminishes as the rate of shear increases [see figure 87(b)]. This is known as thixotropy and often occurs with paints. Brushing reduces the viscosity and the paint flows easily. On the other hand, the paint will not drip easily when left undisturbed.

234   Gases, liquids and solids

Chapter 13
Dielectric properties of matter

In this chapter we shall describe the main dielectric properties of matter and see how far they can be explained in terms of simple atomic models. As we shall see the main mechanism is the polarization of individual atoms and molecules by an electrostatic field.

13·1 Basic dielectric relations

13·1·1 *Dielectric constant ε and polarization P*

Suppose we consider a parallel plate condenser carrying a charge $\dfrac{\sigma}{\text{unit area}}$.

The field in free space between the plates [see figure 88(*a*)] is

$$E_v = 4\pi\sigma. \tag{13.1}$$

The potential difference between the plates is $4\pi\sigma d$ and the capacity C per sq. cm. of condenser is

$$C = \frac{\text{Charge}}{\text{Potential}} = \frac{\sigma}{4\pi\sigma d} = \frac{1}{4\pi d}. \tag{13.2}$$

We now consider what happens if we fill the intervening space with a dielectric. We find that the capacity of the condenser is increased by a factor ε. This is known as the dielectric constant of the material.

For an *isolated* condenser, since σ remains unchanged, this means that the potential difference between the plates is reduced by a factor ε, that is the field is reduced from $E_v = 4\pi\sigma$ to

$$E = \frac{4\pi\sigma}{\varepsilon}. \tag{13.3}$$

This reduction in field is due to the polarization of the material. The field distorts the atoms and molecules, and orients existing dipoles in such a way as to create a field opposing the applied field. The resultant field is thus reduced.* We define a polarization P as the electrostatic moment per unit

---

* If we had considered a condenser maintained in a circuit at a constant potential we could show that the insertion of the dielectric involves a further flow of charge such that the total charge, for the same potential, is increased by a factor ε. Here again the free charges on the condenser plates are partly offset by the induced charges in the dielectric.

volume induced in the dielectric. As we shall see later P is very nearly proportional to the resultant field E. Consider a bar of dielectric of section A, length l. Its total moment due to polarization is

$$P \times \text{volume} = PAl = PA \times l = \text{end charge} \times \text{length}.$$

Thus the moment per unit volume is equivalent to a surface charge P per unit area. We now see that the effective charge density on the plates is $\sigma - P$ [see figure 88(b)]. The resultant field is then

$$E = 4\pi(\sigma - P) = 4\pi\sigma - 4\pi P. \tag{13.4}$$

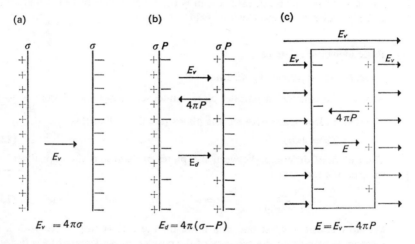

Figure 88. (a) The field in an isolated condenser is E_v. (b) When filled with a dielectric the material is polarized and the field is reduced to E. (c) Relation between E and E_v.

But from equation (13.3), $\varepsilon E = 4\pi\sigma$. Equation (13.4) becomes

$$E = \varepsilon E - 4\pi P$$

$$P = \frac{\varepsilon - 1}{4\pi} E. \tag{13.5}$$

Thus a dielectric which reduces the electrostatic field by a factor ε produces a polarization P which is $\dfrac{\varepsilon - 1}{4\pi}$ times the real field. In what follows we shall show how P can be derived in terms of atomic models. Before doing so, however, we refer for completeness to the displacement vector D.

236   Gases, liquids and solids

13·1·2 *Displacement vector D*

In passing from vacuum to a dielectric the field changes from E_v to E. Taking the simplest case of a parallel block of dielectric normal to the field we see from figure 86(c) that

$$E = E_v - 4\pi P,$$

or $\qquad E_v = E + 4\pi P = E\varepsilon.$

There is a discontinuity in the field. Another quantity called the displacement vector D may now be defined

$$D = \varepsilon E. \tag{13.6}$$

Then in vacuum, $D_v = \varepsilon_v E_v = E_v$ (since $\varepsilon_v = 1$), and in the dielectric

$$D = \varepsilon E = E_v. \tag{13.7}$$

The quantity D is thus continuous. We shall not use it further.

13·2 **Polarization of gases**

13·2·1 *Mechanisms of polarization*

We first consider the behaviour of gases since the molecules are, generally, so far apart that interactions between them may be neglected. Consequently each gas molecule 'sees' the dielectric field E. This polarizes the molecule in three ways:

  (i) By distorting the electronic cloud around the nucleus or nuclei of each molecule. We call the polarization due to this mechanism P_e.
 (ii) By stretching or bending bonds in polar molecules; we call this P_a Usually this is small ($P_a \sim 0.1 P_e$) and can be neglected.
(iii) By orienting existing molecular dipoles: we call this P_d.

We first consider the electronic polarization P_e. Suppose the molecule is initially non-polar. As a result of electronic displacement the individual molecule (or atom) will acquire a dipole moment p. As we shall see immediately p is proportional to E, so we may write

$$p = \alpha E, \tag{13.8}$$

where α is the molecular (or atomic) polarizability. If there are n molecules per cm.^{3},

$$P_e = np = n\alpha E. \tag{13.9}$$

We at once note that the dimensions of P_e are $\dfrac{\text{charge}}{\text{area}}$ whilst the dimensions of E are $\dfrac{\text{charge}}{(\text{distance})^2}$, so they are dimensionally identical. Hence $n\alpha$ has zero

237 Dielectric properties of matter

dimensions or

$$[\alpha] = \left[\frac{1}{n}\right] = [L]^3.$$

In fact as we shall see in what follows α is equal to (molecular radius)³.

13·2·2   *The polarization of neutral monatomic molecules*

Consider a neutral atom such as argon or neon, which consists of a central positive charge Ze and an electron cloud of charge $-Ze$ [see figure 89(a)]. We assume that the electron cloud has uniform charge density ρ and extends over a sphere of radius r. If a field E is applied the electron cloud experiences a force EZe in one direction and the nucleus an equal force in the opposite direction. The charges are displaced relative to one another until the attractive force between them exactly equals the force exerted by the field. Let the displacement be x [see figure 89(b)]. The attraction between the nucleus and

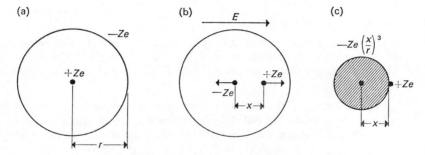

Figure 89.   An atom of atomic number Z considered as a central positive charge surrounded by a negative charge of uniform density. The application of a field E displaces the positive relative to the negative charge.

the electron cloud is the force between Ze on the nucleus and that part of the electronic charge which is within the sphere of radius x.

The latter is $Ze\left(\dfrac{x}{r}\right)^3$. The attractive force [see figure 89(c)] is then

$$f = \frac{\text{(positive charge)} \times \text{(negative charge)}}{x^2}$$

$$f = Ze \times Ze \left(\frac{x}{r}\right)^3 \frac{1}{x^2} = \frac{Z^2 e^2 x}{r^3}.\tag{13.10}$$

This is balanced by the distorting force

$$f = EZe. \qquad (13.11)$$

Equating (13.10) and (13.11) we have

$$E = \frac{Zex}{r^3}. \qquad (13.12)$$

But Zex is the charge times the distance between the charges, i.e., it is the atomic dipole p. Hence

$$p = r^3 E. \qquad (13.13)$$

A similar result is obtained if the atom is assumed to be a perfectly conducting sphere of radius r. Hence

$$P_e = nr^3 E.$$

Combining this with equation (13.5) we have

$$nr^3 E = \frac{(\varepsilon - 1)}{4\pi} E,$$

or $\qquad \varepsilon - 1 = 4\pi nr^3. \qquad (13.14)$

13·2·3 *Dielectric properties and van der Waals critical data*

We now make a wide-ranging and rather unexpected correlation between ε and critical gas data. We note that $4\pi nr^3$ is three times the volume of the molecules in 1 cm.^{3} of gas. The critical volume of the gas is 12 times the volume of the molecules. It follows that $4\pi nr^3$ is equal to $\frac{1}{4}\frac{\rho}{\rho_c}$ where ρ is the density of the gas and ρ_c the density at the critical point. Hence

$$(\varepsilon - 1) = \frac{1}{4}\frac{\rho}{\rho_c}. \qquad (13.15)$$

According to Maxwell's theory of electromagnetic waves, at the same wavelength, the refractive index $\mathscr{R}$ is related to the dielectric constant ε and the magnetic permeability μ by the relation

$$\mathscr{R} = \sqrt{(\varepsilon\mu)}. \qquad (13.16)$$

For dielectrics μ is equal to unity to within one part in 10^6. Since for gases ε is only slightly greater than unity we may write

$$\mathscr{R} = \sqrt{\varepsilon} = \sqrt{[1 + (\varepsilon - 1)]} = 1 + \tfrac{1}{2}(\varepsilon - 1), \qquad (13.17)$$

so that $\quad (\mathscr{R} - 1) \simeq \tfrac{1}{2}(\varepsilon - 1). \qquad (13.18)$

Combining with (13.15) we have

$$\frac{1}{4}\left(\frac{\rho}{\rho_c}\right) = (\varepsilon - 1) = 2(\mathscr{R} - 1). \tag{13.19}$$

These quantities have been compared in Table 27 for inert (monatomic) gases, non-polar polyatomic gases, and for molecules with dipoles. The dielectric constant is found from static or low-frequency measurements.

Table 27 *Dielectric properties of gases*

1	2	3	4	5	6	7
Type	Gas	$\rho \times 10^3$ g. cm.$^{-3}$	ρ_c g. cm.$^{-3}$	$\frac{1}{4}\frac{\rho}{\rho_c}$ $\times 10^4$	$(\varepsilon - 1)$ $\times 10^4$	$2(\mathscr{R} - 1)$ $\times 10^4$
Inert	He	0·18	0·069	6·5	0·7	0·7
	Ne	0·90	0·48	4·7	12·7	13·4
	Ar	1·78	0·53	8·4	5·5	5·7
	Xe	5·90	1·15	12·2	15·0	14·0
Non-polar	H_2	0·09	0·031	7·2	2·7	2·8
molecules	N_2	1·25	0·31	10·0	5·8	5·9
	O_2	1·43	0·43	8·3	5·2	5·4
	Cl_2	3·21	0·57	14·1	—	15·4
	CO_2	1·98	0·46	10·8	9·9	9·9
Molecules	NH_3	0·77	0·23	8·4	71	7·5
with dipoles	SO_2	2·93	0·63	11·6	82	13·2
	H_2O	0·60	0·32	4·7	80	5·0

There are three clear conclusions. First, for the inert and non-polar molecules the agreement between $\frac{1}{4}\frac{\rho}{\rho_c}$ and the observed values of $(\varepsilon - 1)$, see columns 5 and 6, is reasonably good except for He where the inner two electrons are very tightly bound. We conclude that our calculation of the molecular polarization is valid for polyatomic molecules as well as for monatomic molecules. Secondly, for all these gases there is very good agreement between $(\varepsilon-1)$ and $2(\mathscr{R}-1)$ – columns 6 and 7. Since the optical refractive index is a measure of the interaction of the molecules with the high-frequency electric field associated with light waves we deduce that this is the same as the electronic polarization produced by a static or low-frequency electric field in these cases.

Thirdly we note that polar molecules give good agreement between $\dfrac{1}{4}\dfrac{\rho}{\rho_c}$ and $2(\mathscr{R}-1)$ – columns 5 and 7 – since both are a measure of high-frequency electronic polarization. There is poor agreement between $(\varepsilon-1)$ and $2(\mathscr{R}-1)$ – columns 6 and 7 – since ε is a low-frequency determination and includes the orientation effects of the molecule. By contrast $\mathscr{R}$ involves optical frequencies where only the electronic polarizations can respond. Indeed the difference between $(\varepsilon-1)$ and $2(\mathscr{R}-1)$ for ammonia suggests that the orientation of the dipole contributes 63×10^{-4} e.s.u. to the low-frequency dielectric constant.

13·3   Polarization of polar molecules

13·3·1   Apart from the electronic polarization, polar molecules show two other effects:

(i) The stretching or bending of bonds. This produces a polarization P_a which is generally very small. We shall not attempt to estimate it quantitatively.

(ii) An orientation of existing dipoles. The polarization P_d due to this is large and may be as much as $100\,P_e$.

Materials in which existing dipoles are oriented by the applied field are sometimes referred to as para-electric materials. In what follows we provide a simplified model by means of which we can estimate the magnitude of P_d quantitatively.

The simplest example of a dipole (see figure 90) is a diatomic molecule

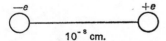

Figure 90.   A simple electrostatic dipole.

with a charge e at one end and $-e$ at the other and a separation of say 1 Å. The permanent dipole has a moment

$$p = \text{charge} \times \text{separation}$$
$$= 4\cdot8 \times 10^{-10} \times 10^{-8}$$
$$= 4\cdot8 \times 10^{-18} \text{ e.s.u.} \tag{13.20}$$

In the absence of a field these dipoles are randomly oriented and show no resultant moment. If an electric field could orient all the molecules we should obtain a polarization per cm.^{3} given by

$$P_d = np.$$

For a gas at s.t.p. n is about 3×10^{19}. Hence

$$P_d \simeq 140 \text{ e.s.u.}$$

The field E_d produced in the dielectric by these dipoles is

$$E_d = 4\pi P_d = 1700 \text{ e.s.u.} = 5 \times 10^5 \text{ volt cm.}^{-1} \tag{13.21}$$

This would be more than enough to ionize most gases, but experiments show that it is virtually impossible to ionize a gas in this way. The reason is that the calculation ignores thermal motion which vigorously opposes the lining-up of the dipoles. In most practical situations the polarization achieved is very much less than that calculated above.

13·3·2 *Orientation of dipoles in an electric field: the Langevin function (para-electric behaviour)*

We consider the effect of a field E on an assembly of dipoles of moment p. The field attempts to orient the dipoles, thermal motion to maintain disorder.

We first calculate the energy of a single dipole in the field if its axis makes an angle θ to the direction of E [see figure 91(a)]. Let the dipole consist of

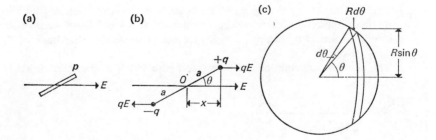

Figure 91. The orientation of a dipole of moment $2aq$ in a field of strength E.

charges q and $-q$ separated by a distance $2a$. Then $p = 2aq$. The potential energy is defined as work done on the system by external forces. The forces on the charges are qE and $-qE$ [figure 91(b)]. If both charges were at some arbitrary position, say at O, we should do work $-qEx$ on the positive charge and $qE(-x)$ on the negative charge in moving them to their present positions.

The total potential energy ε is

$$\varepsilon = -2qEx = -2qEa \cos \theta$$

or $$\varepsilon = -Ep \cos \theta. \qquad (13.22)$$

We now consider how much space is occupied by orientations between θ and $\theta + d\theta$. Construct a sphere of arbitrary radius R [figure 91(c)]. The solid angle subtended between θ and $\theta + d\theta$ is

$$d\omega = \frac{\text{area of annulus}}{R^2} = \frac{2\pi R \sin \theta \, Rd\theta}{R^2},$$

or $$d\omega = 2\pi \sin \theta \, d\theta. \qquad (13.23)$$

We now apply the Boltzmann distribution. The number of dipoles with axes between θ and $\theta + d\theta$ is

$$dn = B \exp \left(\frac{-\varepsilon}{kT} \right) \times (\text{element of space})$$

$$= B \exp \left(\frac{pE \cos \theta}{kT} \right) 2\pi \sin \theta \, d\theta. \qquad (13.24)$$

As this is a rather clumsy expression we make the following substitutions:

$$\left. \begin{array}{l} x = \dfrac{pE}{kT} \\ z = \cos \theta \text{ so that } dz = -\sin \theta \, d\theta \end{array} \right\}. \qquad (13.25)$$

Then equation (13.24) becomes

$$dn = -2\pi B \exp (zx) \, dz. \qquad (13.26)$$

We find B by integrating dn for θ from 0 to π, that is for z from $+1$ to -1. This gives n, the total number of dipoles per cm.[3]

$$n = 2\pi B \int_{-1}^{1} \exp (zx) \, dz. \qquad (13.27)$$

(The change in sign for n follows because we have reversed the order of the limits of integration.) Because of symmetry the resultant polarization normal to E will be zero, and only the polarization parallel to E will be additive. Each dipole contributes a dipole moment parallel to E of amount $p \cos \theta$. Thus the resultant dipole moment per unit volume P_d will be

$$P_d = \int_{\theta=0}^{\theta=\pi} p \cos \theta \, dn$$

$$= 2\pi Bp \int_{-1}^{+1} z \exp (zx) \, dz. \qquad (13.28)$$

243   Dielectric properties of matter

We can express the ratio of P_d to n

$$\frac{P_d}{n} = \frac{\text{equation (13.28)}}{\text{equation (13.27)}} = \frac{p\int ze^{zx}dz}{\int e^{zx}dz}, \tag{13.29}$$

where the integral is from $z = -1$ to $z = +1$.

In most practical situations zx is small so that

$$e^{zx} \simeq 1 + zx. \tag{13.30}$$

Equation (13.29) then becomes

$$\frac{P_d}{n} = \left[\frac{p\int z(1+zx)dz}{\int(1+zx)dz}\right]_{-1}^{+1} = p\frac{x}{3}. \tag{13.31}$$

Substituting from equation (13.25) for x we have

$$P_d = \frac{np^2E}{3kT}. \tag{13.32}$$

We may at once note that for most dielectrics E cannot exceed about 10^5 volt cm.$^{-1} \simeq 300$ e.s.u., otherwise breakdown occurs. For a typical value of $p = 4\cdot8 \times 10^{-18}$ e.s.u., using $k = 1\cdot4 \times 10^{-16}$, and $T = 300°$K., we have

$$\frac{pE}{kT} \simeq 0\cdot03,$$

so that our assumption implied in equation (13.30) is valid.

Exact integration of equation (13.29) gives the Langevin function

$$P_d = np\left[\coth\left(\frac{pE}{kT}\right) - \frac{kT}{pE}\right]. \tag{13.33}$$

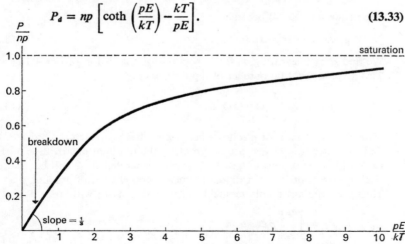

Figure 92. The Langevin relation for the polarization of dipoles produced by a field E.

This function is plotted in figure 92 and shows that under normal conditions the linear part of the curve, far removed from saturation, is the only part that is observable. The contribution of P_d to the dielectric constant may now be calculated. We have

$$P_d = \frac{(\varepsilon-1)_d}{4\pi} E = \frac{np^2 E}{3kT}$$

or $\qquad (\varepsilon-1)_d = \frac{4\pi np^2}{3kT}.$ (13.34)

For ammonia at s.t.p. we have $n = 3 \times 10^{19}$, $p \simeq 1 \cdot 5 \times 10^{-18}$ e.s.u. Then

$$(\varepsilon-1)_d = 67 \times 10^{-4} \text{ e.s.u.}$$

This agrees extremely well with the value quoted in the discussion on page 241 following Table 27. Measurements of the dielectric constant are often used by the chemist to deduce the value of the molecular dipole.

The orientation polarization can, of course, be increased without exceeding the breakdown of the dielectric by reducing the temperature. This cannot, however, be carried very far since most *polar* substances solidify fairly readily and under these conditions molecular orientation becomes very difficult or impossible.

13·3·3 *Effect of frequency of electric field*

We can at once see the powerful effect of frequency on the polarization. We write

$$P = P_d + P_a + P_e$$

$$= \frac{np^2}{3kT} E + P_a + nr^3 E.$$

In a static or low-frequency field all these processes will exert their full part. However, the orientation of the whole molecule is relatively slow, and as the frequency increases the orientation lags behind. When the frequency reaches a value of about 10^{12} the dipoles are unable to follow the oscillations of the field: only random orientations remain and these contribute nothing to the resultant polarization. Of the polarization, P, only P_a and P_e remain. At a somewhat higher frequency the stretching of the bonds becomes too sluggish and P_a drops out. Only P_e remains above a frequency of 10^{15}. This is in the optical range. The behaviour is sketched in figure 93. However, this ignores the fact that both P_a and P_e may show resonance effects. In the next section we shall consider only P_e and show that resonance occurs at a critical frequency and that this accounts for optical dispersion and anomalous dispersion.

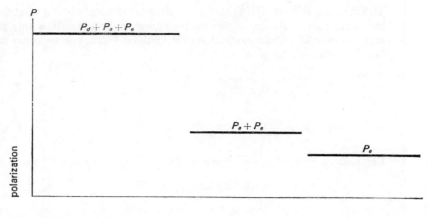

Figure 93. Effect of frequency on polarization ignoring the effects of resonance.

13·4   Optical dispersion and anomalous dispersion

We now analyse the effect of an alternating electric field on the electronic polarization P_e; the treatment is similar to that given by G. Joos. For simplicity we consider a single electron, mass m, charge $-e$, bound to its equilibrium position relative to its nucleus by a force of force constant k. In free oscillation the equation of motion of the electron is

$$m\ddot{x} + kx = 0,$$

or
$$\ddot{x} + \omega_0^2 x = 0, \tag{13.35}$$

where ω_0 is the natural angular frequency (radians sec.$^{-1}$) and $\omega_0^2 = k/m$. The natural frequency ν_0 is given by

$$\nu_0 = \frac{\omega_0}{2\pi} \text{ cycles sec.}^{-1} \text{ (Hertz)}. \tag{13.36}$$

We apply a sinusoidal electric field

$$E = E_0 \sin \omega t, \text{ frequency } \nu = \frac{\omega}{2\pi}. \tag{13.37}$$

Ignoring any 'frictional' effects the equation of motion is

$$\ddot{x} + \omega_0^2 x = \frac{-e}{m} E \sin \omega t. \tag{13.38}$$

The steady-state solution is

$$x = \frac{-eE_0 \sin \omega t}{m} \frac{1}{\omega_0^2 - \omega^2}.$$ (13.39)

The dipole moment per atom is simply

$$p = -ex.$$ (13.40)

Hence

$$p = \frac{e^2 E_0 \sin \omega t}{m(\omega_0^2 - \omega^2)} = \frac{e^2}{4\pi^2 m(v_0^2 - v^2)} E.$$ (13.41)

The volume polarization P_e is simply n times this. Using the relationship

$$(\varepsilon - 1) = 2(\mathscr{R} - 1) = 4\pi \frac{P}{E} = \frac{4\pi np}{E},$$

and substituting for p from equation (13.41) we obtain

$$\mathscr{R} = 1 + \frac{ne^2}{2\pi m(v_0^2 - v^2)}.$$ (13.42)

This is plotted in figure 94(a). It is seen that approaching v_0 the refractive

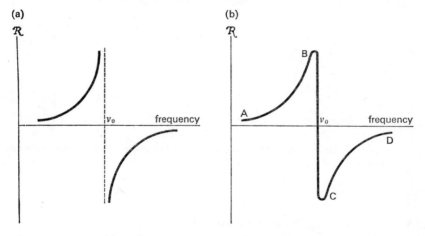

Figure 94. (a) For an undamped system, resonance at frequency v_0 produces a change in $\mathscr{R}$ from $+\infty$ to $-\infty$. (b) With a damped system the transition is sharp but the infinite values do not occur.

index rises to plus infinity and just beyond v_0 it falls to minus infinity. This is the absorption frequency of the atom and is generally of the order of 10^{15}. In practice radiation resulting from the oscillation involves a certain amount of damping or 'friction', and this prevents infinite values of $\mathscr{R}$ from being reached. The main features are now shown in figure 94(*b*). The refractive index increases with frequency over the region AB; this is the region of 'normal' dispersion. The sudden change in behaviour along CD was, at one stage, considered anomalous and was so called. The above analysis shows, however, that 'anomalous' dispersion is just as normal as 'normal' dispersion.

The frequency ν_0 at which electronic resonance occurs is the same as that involved in the derivation of van der Waals forces. For this reason these forces are sometimes called 'dispersion' forces.

Similarly the frequency at which a resonance effect occurs in P_a is of the order of 10^{13}, so dispersion also occurs in the infrared region. The combined behaviour for a dipolar molecule with a single electronic and atomic absorption line is shown schematically in figure 95.

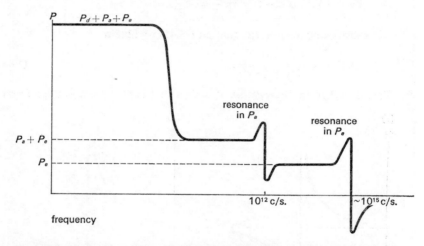

Figure 95. Effect of frequency on polarization of a dielectric allowing for resonance effects.

13·5 Dielectric properties of liquids and solids

13·5·1 *Weak interaction*

We now consider the dielectric behaviour of matter in the condensed state. We first treat the simplest case in which non-polar molecules are involved so that interaction between neighbouring atoms or molecules is weak. In this case we consider only the electronic polarization per atom

$$p = r^3 E, \tag{13.43}$$

where r is the radius of the individual atom. For simplicity, to obtain an order of magnitude value for the number of atoms per c.c., n, assume a simple cubic structure.

Then

$$n = \left(\frac{1}{2r}\right)^3. \tag{13.44}$$

The polarization per unit volume, using equations (13.43) and (13.44), is

$$P = np = \frac{E}{8}. \tag{13.45}$$

Hence

$$\varepsilon - 1 = \frac{4\pi P}{E} = \frac{\pi}{2}. \tag{13.46}$$

Consequently

$$\varepsilon = 1 + \frac{\pi}{2} \simeq 2\cdot6,$$

$$\mathscr{R} = \sqrt\varepsilon \simeq 1\cdot6. \tag{13.47}$$

This analysis shows that for many simple liquids and solids the dielectric constant will be of order 2–3 and the refractive index 1·4 to 1·7. Clearly in a crystal which has well defined structural anisotropy, so that the polarizability depends on crystal direction, there will be a corresponding dependence of ε and $\mathscr{R}$ on crystal direction. This is the basic cause of birefringence.

13·5·2 *Strong interaction: the Clausius–Mosotti equation*

When atoms or molecules are far apart as in a gas, they 'see' the average dielectric field E. When, however, they are close together as in a liquid or solid, the atomic (or molecular) dipoles may themselves interact. For example if polarization produces a local dipole of strength $p = r^3E$, the field produced by this dipole at a distance r is of order of magnitude p/r^3, i.e., about E; that is, the field produced at a neighbour by an induced dipole may be about as large as the field itself. This increases the polarization effect.

We have now to consider how to estimate the resultant field acting on each molecule. The treatment is due to Lorentz. We choose a molecule O and construct a sphere of radius R large compared with the molecular separation (see figure 96). The resultant effect at O is the sum of (*a*) the field produced by the material within R and (*b*) the field produced by the material outside R. The latter can be treated as a continuum since it is so far away from O. The field due to this is the same as the equivalent surface charges over the spherical surface. The calculation is as follows.

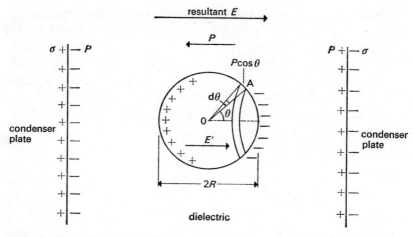

Figure 96.　Charge distribution in a spherical cavity within a dielectric.

Construct an annulus on the surface of the cavity making an angle θ, $\theta + d\theta$ with the direction of the polarization field. The surface charge density is

$$-P \cos \theta. \tag{13.48}$$

The charge on the annulus is

$$-P \cos \theta \, 2\pi R \sin \theta \, R d\theta. \tag{13.49}$$

This produces a coulomb field at O pointing along OA. The vertical components cancel out, the horizontal ones are additive. The horizontal component at O, for the annulus, is

$$P \cos \theta \, 2\pi R \sin \theta \, R d\theta \, \frac{1}{R^2} \cos \theta$$

$$= 2\pi P \cos^2 \theta \sin \theta \, d\theta. \tag{13.50}$$

For the whole sphere, for θ ranging from 0 to π, the net field E' at O due to the surface charge is

$$E' = 2\pi P \int_0^\pi \cos^2 \theta \sin \theta \, d\theta = 2\pi P \left[\frac{-\cos^3 \theta}{3} \right]_0^\pi = \frac{4\pi P}{3}. \tag{13.51}$$

The resultant field at O is thus $E + E'$ or

$$E + \frac{4\pi P}{3}. \tag{13.52}$$

We now have to consider the contribution (a) due to material within the sphere. Lorentz showed that for a cubic array of dipoles or for a random array of dipoles as in a liquid or a glass, the local field is zero. For such materials we may, therefore, write

$$E_{local} = E + \frac{4\pi P}{3}.$$ (13.53)

The induced dipole per molecule is

$$p = \alpha E_{local},$$ (13.54)

$$\therefore P = np = n\alpha \left[E + \frac{4\pi P}{3} \right].$$ (13.55)

Hence
$$\frac{P}{E} = \frac{n\alpha}{1 - \frac{4}{3}\pi n\alpha}.$$ (13.56)

The basic relation between P and the average field E in the dielectric is

$$(\varepsilon - 1) = \frac{4\pi P}{E} = \frac{4\pi n\alpha}{1 - \frac{4}{3}\pi n\alpha}.$$ (13.57)

Hence $\quad (\varepsilon + 2) = (\varepsilon - 1) + 3 = \dfrac{3}{1 - \frac{4}{3}\pi n\alpha}.$ (13.58)

Taking the ratio of equations (13.57) and (13.58), we have finally

$$\frac{\varepsilon - 1}{\varepsilon + 2} = \frac{4\pi n\alpha}{3}.$$ (13.59)

This relation is fairly well obeyed for polar liquids and for dilute solutions of dipolar molecules in a non-polar solvent, as well as for ionic solids of cubic structure. It is known as the Clausius–Mosotti equation.

Before we leave this section we may form some idea of the importance of interaction compared with the results obtained if interaction is ignored. In the latter case, as we saw above,

$$\varepsilon - 1 = 4\pi \frac{P}{E} = 4\pi n\alpha.$$ (13.60)

Consider three cases:

(a) Weak interaction, say $\varepsilon \simeq 1$; we see that equation (13.60) is identical with equation (13.59).

(b) Slight interaction, say $\varepsilon = 3$; equation (13.59) then gives a value for $(\varepsilon - 1)$, which is $\frac{5}{3}$ times bigger than that given by equation (13.60).

(c) Large dipole interaction, say $\varepsilon = 70$; equation (13.59) gives a value for $(\varepsilon - 1)$ which is 24 times larger than that given by equation (13.60).

There is yet a further type of interaction which we have not discussed here, but shall refer to briefly in the next chapter. This is the analogue of ferromagnetism where interaction is intrinsically so large that even without the presence of an applied field the material is strongly polarized. It would correspond to a situation [see equation (13.58)] where $\frac{4}{3}\pi n\alpha \simeq 1$ so that $\varepsilon \to \infty$. Such materials are known as ferro-electrics. A ferro-electric material such as barium titanate has a dielectric constant of over 1000.

13·5·3   *Effect of frequency*

With matter in the condensed state frequency has the same effect on the dielectric constant as it has on gases. However the dipole polarization in liquids is usually more sluggish than in gases – with solids it is often completely absent since rotation of the whole molecular unit within the lattice is difficult or impossible particularly at temperatures well below the melting point. Results for water and ice are shown in figure 97.

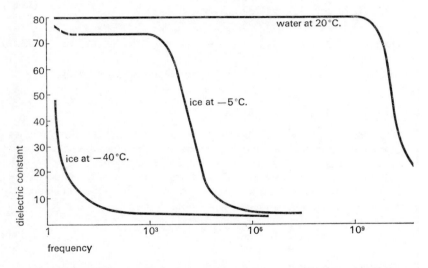

Figure 97. Variation of dielectric constant with frequency, for water at 20°C. and for ice at −5°C. and −40°C. The large dielectric constant at low frequences at −40° C. shows that molecular rotation is still possible (see p. 178).

13·6   **Summary**

We may summarize the ideas described in this chapter in the following way.

1. The dielectric properties of matter arise from the polarization produced by an electrostatic field. Polarization occurs in three ways:

(*a*) Distortion of electron clouds; all atoms (or molecules) show this effect.

(*b*) With polar molecules stretching or bending of bonds can produce a small contribution to the polarization.

(*c*) With polar molecules orientation of the molecule may produce a very large polarization: the field has to compete with thermal randomization.

2. Temperature has no effect on (*a*) since this involves electronic properties. Increasing temperature resists the polarizing influence of the field for both (*b*) and (*c*). It is particularly marked with (*c*). However, the molecular orientation cannot be increased indefinitely by lowering the temperature since orientation of the molecular dipoles may become impossible when the material is solid.

3. Frequency of the electrostatic field has a marked effect on all types of polarization. Orientation becomes increasingly difficult as the frequency is raised and becomes impossible at frequencies above about 10^{12} cycles sec.$^{-1}$ The stretching or bending of bonds also becomes unable to follow the field at frequencies above about 10^{14} cycles sec.$^{-1}$, and only the electronic polarization remains up to a frequency of 10^{15}–10^{16} cycles per second. This is in the optical range and is therefore the only part of the dielectric mechanism which plays a part in the refractive index of the material. There is usually a resonance effect at frequencies in this region and this accounts for optical dispersion. The bond stretching (or bending) mechanism also shows a resonance effect but this generally occurs in the infrared.

4. With matter in the condensed phase the behaviour is complicated by neighbour–neighbour interaction. This can greatly increase the effective polarization of the material.

Chapter 14
Magnetic properties of matter

Magnetic properties are more difficult to interpret than dielectric properties. This is partly because the basic mechanisms are more involved and partly because the historic development of the subject has led to confusion, particularly between the field H and the magnetic induction B. Fortunately the difference between H and B is not of great importance in explaining the mechanism of diamagnetism and paramagnetism. Nevertheless we shall attempt to keep the issue clear.

14·1 The magnetic equations

14·1·1 Field, permeability, magnetic induction and susceptibility

We begin with the concept of a magnetic field, which should be defined in terms of the couple exerted by the field on a small loop carrying a specified current. This is not very practical and it is more convenient to think in terms of the force exerted on a magnetic pole. Isolated poles do not exist but long thin magnets with spherical end-pieces behave as though the poles were isolated at each end. Such magnets may be used to define a unit magnetic pole: two unit poles, one centimetre apart in air, exert a force on one another of 1 dyne. We can then define a magnetic field H in air as the field which exerts a force mH on a pole of strength m. We may wind a coil and find that for a specified current it exerts a force mH on the same pole in air. The field due to the current is then defined as H.

If in this coil we insert a piece of magnetic material the field is increased. The pole of strength m, if placed near the magnetic material, will experience a force greater than mH. We may approach this result from a different angle. Suppose the coil originally possessed self-inductance L. This arises from the change in magnetic flux cutting the turns of the coil when the current is changed. Subsidiary experiments with the coil in air show that the flux is proportional to the magnetic field H which it produces. Suppose now the whole of the space in the coil is filled with magnetic material. It is then found that the self-inductance is increased. Suppose the factor of increase is μ; then μ is defined as the magnetic permeability of the core material. Clearly the magnetic flux has been increased by this factor. We now invent a new quantity B known as the magnetic induction and say that, in the magnetic core,

$$B = \mu H. \tag{14.1}$$

It may be shown that B is continuous across a normal interface exactly as for D. The force exerted on a unit pole outside the core will be increased but this increase will depend on the shape of the end-piece of the core.

The clearest and least problematical situation is that in which the coil is in the form of a torus (see figure 98). If the space is filled with magnetic material

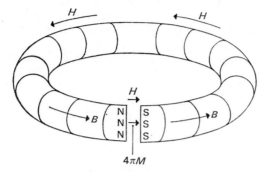

Figure 98.   Effect of magnetic field on magnetic material.

there are no free ends. The increase in magnetic flux is due to polarization of the material imparting to it magnetic moment per unit volume M. The effect of this on the magnetic flux may be considered by cutting a very thin slice out of the core normal to the direction of the field. In this gap the original field H is augmented by M. This is equivalent to a distribution of magnetic poles over the free surface of amount M per sq. cm. (in this case it is a fictitious equivalence since free poles do not exist). The additional field produced is $4\pi M$ so in the gap the field is $H+4\pi M$. Since B is continuous across the interface and μ in free space is unity,

$$B = H+4\pi M. \tag{14.2}$$

This is closely analogous to the dielectric relation given in equation (13.7),

$$D = E+4\pi P,$$

and the similarity has been generally stressed. It is, however, misleading for the following reasons. First, consider the forces experienced by a charge or a magnetic pole within a medium. Within a dielectric the forces on a charge are reduced; within a magnetic material the forces on a pole are increased. This is because the electrostatic forces are determined by (the local values of) E and magnetic forces by B. The latter point has been proved in a direct way by F. Rasetti in 1944 in his study of the deflection of cosmic rays through magnetized iron (F. Rasetti, *Physics Review*, vol. 66, 1944, p. 1). For fast rays, where the cosmic particle sees the material as a continuum, the deflection is found to correspond to B and not to H. This is consistent with the view that magnetized particles behave like circulating currents (see below).

A second difference concerns the direction of the field. If electric charges are placed on a dielectric the direction of the electrostatic field is always from positive to negative whether it is observed inside or outside the dielectric. With a permanent magnet, however, the situation is different: outside the material the field is in the direction north to south; within the material the inductive field (the magnetic induction) is from south to north and a unit north pole would experience a force in that direction. Here again the primary magnetic elements behave like circulating currents and the permanent magnet as a whole behaves like a solenoid.

Before returning to the magnetic equations we may note the following further differences between electrostatic and electromagnetic concepts:

(a) there is no such thing as an isolated magnetic pole

(b) there is no magnetic analogue of a condenser

(c) there is no electrostatic analogue of a solenoid.

Although B is of such primary importance we still need to be able to express magnetic phenomena in terms of H. This is largely because most magnetic experiments involve the use of solenoids where H is known in terms of the number of turns and the current. We may eliminate B from equation (14.2) by writing

$$B = \mu H = H + 4\pi M$$

so that

$$(\mu - 1)H = 4\pi M. \tag{14.3}$$

Since the magnetization M is a function of the field H we may write

$$M = \chi H \tag{14.4}$$

where χ is termed the magnetic susceptibility. Then from equations (14.3) and (14.4) we have

$$\mu - 1 = 4\pi \frac{M}{H} = 4\pi\chi. \tag{14.5}$$

Materials may be divided into three main classes according to their values of χ.

(a) Diamagnetic. χ is small and negative. The material is weakly repelled by a strong magnetic field. All substances possess a diamagnetic component.

(b) Paramagnetic. χ is small and positive. Such materials are weakly attracted by strong magnetic fields.

(c) Ferromagnetic. χ is large and positive. The material is strongly attracted by a magnetic field.

In practice χ is roughly constant only for feebly magnetic materials, i.e., for diamagnetic and paramagnetic substances. For ferromagnetic materials χ varies very markedly with the field itself. In what follows we shall derive expressions for χ in terms of atomic models.

14·1·2   *Force exerted on a current by a field*

A current i is defined as the charge per sec. passing through a conductor. If a short length dl of such a conductor is placed in a field of induction B at an angle θ it is found experimentally that it experiences a force

$$Bi \sin \theta \, dl, \tag{14.6}$$

in a direction at right angles to the plane containing B and dl [figure 99(a)].

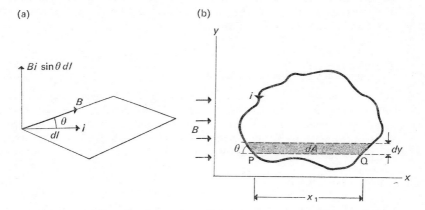

Figure 99.   (*a*) A field *B* acting on a current element *i* produces a force normal to the plane containing *B* and *i*. (*b*) Calculation of moment exerted by a field *B* on a coil carrying current *i*.

The force is determined by B not by H. Consider a coil of irregular shape in the plane xy, the field B pointing in the x direction [figure 99(b)]. There is a force on the element P of amount

$$iB \sin \theta \, dl = iB \, dy, \tag{14.7}$$

out of the paper and an equal force at Q into the paper. These two forces exert a couple

$$iB \, dy \, x_1 = iB \, dA, \tag{14.8}$$

where x_1 is the distance PQ and dA is the area of the strip between P and Q. For the whole coil the couple is then BiA where A is the area of the coil. Thus the coil behaves as though it has a magnetic moment iA so that in a field of induction B it experiences a couple BiA. (This may be extended to a coil which is not planar and leads to the more general concept of the equivalent magnetic shell. As we do not need this in what follows we shall not pursue it further.) Ever since A. M. Ampère, it has been considered that the magnetism of atoms is due to circulating currents within the atom; consequently every north pole has associated with it an equal south pole. For this reason it is not possible to envisage the existence of isolated magnetic poles in the way that there are isolated charges in electrostatics. The earlier view regarded the circulating currents as arising from electrons travelling in orbits; later work has included the electron spin which, from this point of view, can also be regarded as a minute circulating current.

It follows from this discussion that the force on a magnetic dipole depends on B. When, however, we deal with atomic dipoles we must consider the true or local value of B at the atom itself. For paramagnetic and diamagnetic materials, as we shall see, this raises no real problem since μ is extremely small and B and H are almost equal.

14·2   Diamagnetism: Langevin's treatment

All atoms possess a diamagnetic component. This is because an applied field always modifies the circulating 'electronic' current in such a way that it opposes the field. We consider the simplest case, an electron of charge e e.s.u., moving with velocity v in a circular orbit of radius r. If τ is the time for the electron to describe one revolution, e behaves like a current of magnitude

$$i = \frac{e}{\tau} \times \frac{1}{c} = \frac{ev}{2\pi r} \times \frac{1}{c}, \qquad (14.9)$$

where c is the velocity of light and is introduced to express the current in e.m. units.

The magnetic moment of the orbiting electron is

$$p = \pi r^2 i = \frac{evr}{2c}. \qquad (14.10)$$

Consider now the effect of a field of induction B which is normal to the orbit. The flux ϕ in the circuit is

$$\phi = \pi r^2 B. \qquad (14.11)$$

If the field varies with time an e.m.f. will be induced in the circuit of magnitude

$$\text{e.m.f.} = -\frac{1}{c}\frac{d\phi}{dt} = -\frac{\pi r^2}{c}\frac{dB}{dt}. \qquad (14.12)$$

This e.m.f. acts around the current loop and is equivalent to an electrostatic field E, where

$$\text{e.m.f.} = 2\pi r E$$

$$= -\frac{\pi r^2}{c}\frac{dB}{dt}. \qquad (14.13)$$

Hence

$$E = -\frac{r}{2}\times\frac{1}{c}\times\frac{dB}{dt}. \qquad (14.14)$$

This field exerts a force eE on the electron and this in turn produces a change dv in its velocity:

$$eE = m\frac{dv}{dt} = -\frac{re}{2c}\times\frac{dB}{dt}, \qquad (14.15)$$

$$\frac{dv}{dt} = -\frac{re}{2mc}\times\frac{dB}{dt}. \qquad (14.16)$$

Consequently in the time that B changes by ΔB, v changes by Δv where

$$\Delta v = -\frac{re}{2mc}\Delta B. \qquad (14.17)$$

According to equation **(14.10)** a change in v produces a corresponding change in the magnetic moment of the orbit:

$$\Delta p = \frac{er}{2c}\Delta v.$$

Combining with equation **(14.17)**,

$$\Delta p = -\frac{e^2 r^2}{4mc^2}\Delta B. \qquad (14.18)$$

We see that Δp is always in a sense which opposes ΔB. Consider for example two identical orbits with opposite orbital momenta. Each orbit has a magnetic

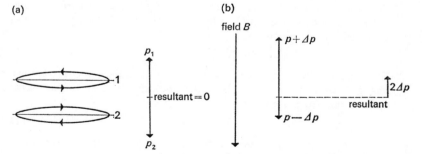

Figure 100.   Sketch showing nature of diamagnetism; (a) when no field is applied the net moment is zero, (b) when field ΔB is applied there is an additive change in the magnetic moments of both orbits.

259   Magnetic properties of matter

moment p but the net moment is zero. If a field ΔB is applied one orbit is slowed down and the other is speeded up: the net change is additive for the two orbits (see figure 100).

There is another point of interest. If the orbital velocity is reduced by Δv, the centrifugal force on the electron is reduced by $\Delta\left(\dfrac{mv^2}{r}\right) = \dfrac{2mv\Delta v.}{r}$ From equation (14.17) this equals $\dfrac{ve\Delta B}{c}$. But the field ΔB itself exerts a radial force on the electron of amount: velocity $\times$ charge $\times$ $\Delta B = \dfrac{ve\Delta B}{c}$. This exactly compensates for the reduced centrifugal force: consequently if the initial orbit is stable under some specified force between nucleus and electron the diamagnetic mechanism does not introduce any change.

For diamagnetic materials $B \simeq H$ to 1 part in 10^5. Equation (14.18) then becomes

$$\Delta p = -\frac{e^2 r^2}{4mc^2}\,\Delta H. \tag{14.19}$$

By definition the magnetic susceptibility per electron is simply $\dfrac{\Delta p}{\Delta H}$. If there are n atoms per unit volume the susceptibility per unit volume is

$$\chi = n\frac{\Delta p}{\Delta H} = -\frac{n}{4}\sum\frac{e^2 r^2}{mc^2}, \tag{14.20}$$

where the summation is for all the orbital electrons in each atom.

If the orbits are not normal to B each orbit undergoes precession and χ is slightly decreased. Similarly if the orbits are not circular a lower value of χ is obtained. A better value for χ is

$$\chi = -\frac{n}{6}\sum\frac{e^2 r^2}{mc^2}. \tag{14.21}$$

We may at once derive an order-of-magnitude value for χ. If a gram-atom occupies 10 c.c., $n = 6 \times 10^{22}$. If there are say 10 orbital electrons and r is of order 10^{-8} cm. we have

$$\chi = -(2 \text{ to } 3) \times 10^{-6}. \tag{14.22}$$

This is a typical value for most solids and liquids. In the gaseous state the main change is in n. At s.t.p. n is about 10^3 times smaller than in the condensed state so that for gases at s.t.p.

$$\chi \simeq 10^{-9}. \tag{14.23}$$

Note that χ is independent of temperature since the internal structure of the atom is virtually unchanged by normal temperatures.

In the model described above, diamagnetism is attributed to the influence of the magnetic field on the orbital momentum of the electrons. What then happens to the s-electrons, which possess zero orbital momentum? A detailed quantum theory treatment shows that s-electrons possess zero angular momentum only in the absence of a magnetic field. Upon the application of a magnetic field they acquire a small amount of orbital momentum. This produces an equivalent amount of diamagnetism to that calculated above. Thus it follows that all the electrons contribute to diamagnetism.

14·3   Paramagnetism: the Langevin function

If an atom possesses a permanent magnetic dipole because the electrons have either orbital momentum or spin, the dipole will orient in an applied magnetic field and produce a magnetization in the same direction as the applied field. Materials which are made of such atoms are attracted (weakly) by a magnetic field and are known as paramagnetic substances. As in the case of dielectric dipoles the orientation by the applied field is opposed by thermal disorder.

We first consider the order of magnitude of the permanent dipole p arising from an orbiting electron. For a circular orbit of radius r and electron velocity v we have [see equation (14.10)]

$$p = \frac{evr}{2c}.$$ (14.24)

But the angular momentum mvr is quantized in units of $\frac{h}{2\pi}$ where h is Planck's constant. Putting

$$mvr = n_0 \left(\frac{h}{2\pi}\right),$$ (14.25)

where n_0 is an integer, and substituting in equation (14.24) we have

$$p = n_0 \left(\frac{eh}{4\pi mc}\right).$$ (14.26)

The quantity in the bracket is known as the Bohr magneton. As a first approximation we expect to find certain atoms in which the angular momenta do not cancel out, so that one orbit (or one spin) is left unpaired. The magnetic moment of such an atom will be one Bohr magneton the magnitude of which is

$$\frac{eh}{4\pi mc} \simeq 10^{-20} \text{ e.m.u.}$$ (14.27)

Since orbits, and spins, tend to occur in pairs of opposite momenta (or spins) we should not expect to find atoms with magnetic moments much greater than this. This has an important consequence which greatly eases further analysis. As we shall see below, at low temperatures and moderate fields H, it is possible to orient all the dipoles and produce magnetic saturation. Nevertheless, even at this stage, the magnetic moment M per unit volume is small compared with H. Consequently the difference between B and H, or between H and the true field seen by the individual atom, is negligible and can be ignored. In what follows, therefore, we consider simply the effect of the applied field H.

Following the analysis for para-electric behaviour given in the previous chapter, we expect to find for small and moderate fields a constant susceptibility given by

$$\chi = \frac{np^2}{3kT}. \tag{14.28}$$

For a solid occupying 10 c.c. per gram-atom, $n = 6 \times 10^{22}$. Assuming p is one magneton we find that, at 300°K., χ has a value

$$\chi \simeq 10^{-4}.$$

This is 50 to 100 times greater than the diamagnetic susceptibility. Consequently the diamagnetic susceptibility, which all atoms possess, is swamped by the paramagnetic susceptibility if the atom contains an unpaired orbit or spin.

For strong fields or low temperatures we obtain the corresponding Langevin function

$$\chi = \frac{np}{H} \left[\coth \frac{pH}{kT} - \frac{kT}{pH} \right]. \tag{14.29}$$

This is drawn in figure 101. Although this curve is identical with that obtained for para-electrics there are two basic differences between paramagnetic and

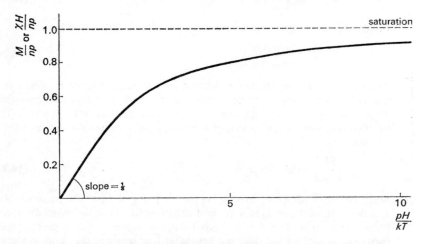

Figure 101. The Langevin function for paramagnetism, treating the orientation as a continuous (classical) phenomenon.

para-electric materials. The first concerns saturation. Low temperatures clearly favour saturation in both cases. However most dielectrics possessing electrostatic dipoles become solids at low temperature and orientation becomes difficult or impossible; orientation is thus frozen-out. With paramagnetics we are only concerned with the orientation of orbits (or spins) and this can occur at the very lowest temperatures. Thus one can greatly enhance paramagnetism by operating at very low temperatures; one cannot in general do this with para-electrics.

As an example of saturation with paramagnetics we note from figure 101 that this occurs when $pH/kT > 1$, i.e., when

$$H > \frac{kT}{p}. \tag{14.30}$$

If $p = 10^{-20}$ e.m.u. and $T = 1°$K. we obtain saturation when

$$H > 1 \cdot 4 \times 10^4 \text{ gauss.}$$

Such a field can be easily obtained and, with it, saturation can be achieved at temperatures of about 1°K.[*]

The second difference concerns the nature of the polarizing field. Consider a typical solid occupying 10 cm.^{3} per gram-atom; then $n \simeq 6 \times 10^{22}$. Suppose all the dipoles are aligned by operating at 1°K. with a field $H = 14,000$ gauss. The magnetic moment per unit volume $M = np \simeq 6 \times 10^2$. The quantity $4\pi M$ still remains appreciably less than H, so B is not greatly different from H. If one constructs a spherical cavity as in the dielectric case the quantity $4\pi M/3$ is less than 20 per cent of H. Thus the whole problem of deciding, 'what is the real field that the individual atom sees?' is relatively unimportant. We may simply assume that it is nearly equal to H without loss of accuracy. Of course if we carried out the experiment at 0·1°K. so that saturation occurred for a field of only 1400 gauss, the problem would become more acute and we should have to carry out an elaborate analysis. In practice, the issue becomes important only in ferromagnetism. In that case, as we shall see below, the interaction due to the magnetic polarization of the material is so strong that it completely dominates the behaviour.

The difference between paramagnetic and para-electric behaviour arises from the size of the unit dipole. In electrostatics it has a value of about $4 \cdot 8 \times 10^{-18}$ e.s.u.; in paramagnetism the Bohr magneton has a value of about 10^{-20} e.m.u. The electrostatic unit dipole is thus about 500 times greater than the corresponding magnetic dipole. For this reason the interaction of the electrostatic dipole with the applied field is far more important.

[*]This treatment is valid for gases and for paramagnetic salts where the individual dipoles play their full part. With paramagnetic metals the major contribution is from the spin of the free electrons and only those at the top of the Fermi level are able to take part, i.e. only a fraction of order kT/E_F, see p. 191.

14·4·1   *Theory of ferromagnetism*

Certain materials such as iron, nickel and cobalt show very high magnetization M (typically $M \simeq 1000$) for fairly modest fields even at room temperature. They reach magnetic saturation very easily. With an 'ordinary' specimen of iron this occurs for fields of the order of a few hundred gauss (see figure 102); with good single crystals in favourable orientations for a field of order only 1 gauss, so under these conditions χ is about 1000.

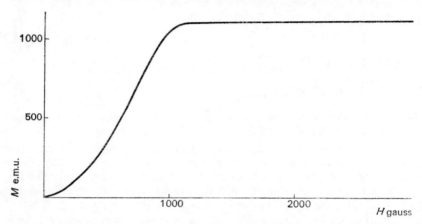

Figure 102. Magnetic moment M produced in a typical magnetic material for fields of order 1000 gauss.

A typical value of M at saturation is about 1000. If a gram-atom occupies 10 c.c. the number of atoms per c.c. is 6×10^{22}. At saturation, if these give a moment of 1000 e.m.u., each atom must contribute a magnetic moment of $\frac{1000}{6 \times 10^{22}} \simeq 1\cdot6 \times 10^{-20}$, i.e., about 1 Bohr magneton. This suggests that a saturated ferromagnetic consists of completely aligned dipoles. In this sense they are like paramagnetics, but with paramagnetics this result can only be achieved by using enormous fields or by operating at extremely low temperatures. With ferromagnetics alignment occurs with relatively weak fields.

It was this conclusion that led A. Weiss to suggest that these materials are spontaneously magnetized over fairly large domains and that these domains are easily aligned when an external field is applied. There is a co-operative interaction between the atoms so that an additional field λM acts on each atom. This resembles the Lorentz field in dielectrics $\left(\frac{4\pi}{3}P\right)$, but the quantity λ is immensely larger, of the order 10^4. The resultant field producing magnetization may then be written

$$H + \lambda M \tag{14.31}$$

(or $B + \lambda'M$ where $\lambda' = \lambda - 4\pi$).

The Langevin equation for the alignment of the dipoles becomes

$$M = np \left[\coth x - \frac{1}{x} \right] \tag{14.32}$$

where now

$$x = \frac{p}{kT} [H + \lambda M]. \tag{14.33}$$

The behaviour of the Langevin equation as a function of H is best seen by rewriting equation (14.32) as

$$\frac{M}{np} = \coth x - \frac{1}{x}, \tag{14.34}$$

and equation (14.33) as

$$\frac{M}{np} = \frac{xkT}{p^2 \lambda n} - \frac{H}{\lambda np}. \tag{14.35}$$

Consider first the behaviour when $H = 0$. This corresponds to the spontaneous magnetization of ferromagnetic materials in the absence of an applied field. We plot equation (14.34) as $\dfrac{M}{np}$ against x and obtain the typical curve of figure 103(a). The slope near the origin is $\frac{1}{3}$. In figure 103(b) we have plotted equation (14.35) as $\dfrac{M}{np}$ against x, for $H = 0$. We obtain

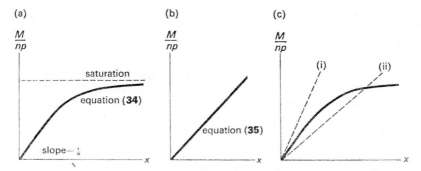

Figure 103. Graphical method of explaining ferromagnetism. (a) Graph of equation (14.34). (b) Graph of equation (14.35). (c) Superposition of both graphs. For curve (i) there is no intersection, i.e. ferromagnetism does not occur. For curve (ii) intersection implies that ferromagnetism does occur.

a straight line of slope $\dfrac{kT}{p^2 \lambda n}$. Clearly if this slope is greater than $\frac{1}{3}$ no solution is possible and there is no magnetization [figure 103(c), curve (i)].

If it is less than $\frac{1}{3}$ magnetization and saturation are easily achieved [figure 103(c), curve (ii)].

It is at once evident that there is a critical temperature above which no ferromagnetization will occur. This is known as the Curie temperature and is given by

$$\frac{kT_c}{np^2 \lambda} = \frac{1}{3}.$$ (14.36)

If the saturation value of M is written as M_0 ($M_0 = np$) the drop-off of M with increasing temperature is shown in figure 104 as a plot of M/M_0 against

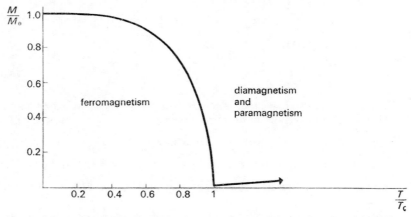

Figure 104.   Drop in magnetic moment of a ferromagnetic material as temperature is raised. Above the Curie temperature T_c the material shows only diamagnetism and paramagnetism.

T. Above T_c only paramagnetism and diamagnetism remain. In what follows we shall ignore the diamagnetism and calculate in a very simple way the paramagnetism above T_c. If the fields are not too strong the Langevin equation (14.34) may be written

$$\frac{M}{np} = \frac{x}{3}.$$ (14.37)

Similarly, using equation (14.36), equation (14.35) may be written

$$\frac{M}{np} = \frac{xT}{3T_c} - \frac{H}{\lambda np}.$$ (14.38)

Substituting for x from equation (14.37) we obtain

$$\frac{M}{np}\left(\frac{T}{T_c}-1\right) = \frac{H}{\lambda np}. \tag{14.39}$$

Consequently the paramagnetic susceptibility becomes

$$\chi = \frac{M}{H} = \frac{1}{\lambda}\left(\frac{T_c}{T-T_c}\right). \tag{14.40}$$

From equation (14.36) this becomes

$$\chi = \frac{np^2}{3k(T-T_c)}. \tag{14.41}$$

We see at once that λ has disappeared explicitly although, of course, it is implicit in the value of T_c. Equation (14.41) is well obeyed and gives good quantitative agreement for most ferromagnetic materials on the assumption that each atom contributes of the order of 1 or 2 Bohr magnetons.

Finally we may note that in the ferromagnetic range an applied field has little effect compared with the internal field. Even for a field $H \simeq 10^4$ gauss the interval $\dfrac{pH}{kT}$ shown in figure 105 is small and has very little effect on the resultant magnetization.

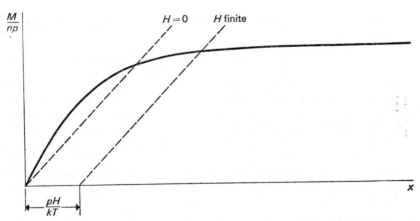

Figure 105. An applied field H has little effect on the magnetization of a ferro-magnetic material.

14·4·2 *The origin of the internal field: ferromagnetism and anti-ferromagnetism*

It is not possible to explain the origin of the large internal field in simple terms. Indeed it introduces very difficult quantum mechanical problems.

We may, however, describe it in the following terms. The large value of λ arises from the 'exchange' interaction between the atomic dipoles. This interaction depends on the ratio of the orbital radius r to the distance a between neighbouring atoms. Over a critical range (a/r about 3) the magnetic exchange interaction is large and positive. This implies that for atoms with large dipoles and of the correct a/r ratio strong positive interaction will occur, λ will be large and the material will be ferromagnetic (Fe, Co, Ni). This is shown schematically in figure 106. It is also seen that for small

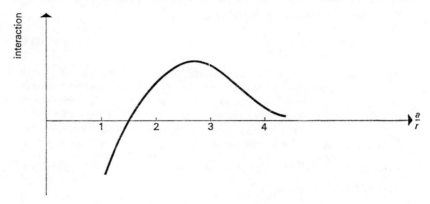

Figure 106. Exchange interaction in ferromagnetic materials as a function of $\frac{a}{r}$.

values of a/r the interaction can be negative. This favours an anti-parallel orientation of neighbouring spins and materials showing this property (MnF_2, MnO, CoO) are known as anti-ferromagnetic substances: at low temperatures the interaction is effective and an applied field produces little magnetization so that the susceptibility is small. As the temperature is raised the anti-ferromagnetic interaction becomes less effective and the susceptibility of the material increases. Above a critical temperature known as the Néel temperature the interaction is completely swamped by thermal motion. The material ceases to be anti-ferromagnetic and becomes paramagnetic in the, more or less, classical sense. The detailed behaviour is very complicated and we shall not discuss it further.

14·5 Quantum treatment of magnetic properties

The classical treatment of diamagnetism is satisfactory from two points of view. First the concept of induced currents producing fields which oppose the applied field seems reasonable and tractable. Secondly it gives the right answer for the diamagnetic susceptibility. From all other points of view it is wrong.

Detailed considerations, first described by N. Bohr in 1911, show that if classical mechanics are applied *rigorously* to an orbiting electron the paramagnetic and diamagnetic terms cancel one another exactly so that there is no net induced magnetization. The reason is briefly as follows. Assuming that an electron orbiting around a nucleus is a stable classical system the radius of a given orbit can have a continuous range of values from 0 to ∞ rather than the one particular size assumed in the Langevin theory. Further, whereas the Langevin theory applies the Boltzmann energy distribution to the external degrees of freedom of the orbiting electron, that is to the orientation of the orbits in paramagnetism, it does not do so to the internal degrees of freedom, that is to the orbital energies of the electron in its treatment of diamagnetism. This is logically inconsistent. If the Boltzmann distribution is applied to both orientation energies and orbital energies the paramagnetic and diamagnetic parts exactly compensate one another. The Langevin treatment gives the correct answer because it assumes a finite radius and an orbital energy independent of temperature; in a sense it incorporates hidden quantum conditions although in fact Langevin's theory (1905) preceded Bohr's quantized atom (1913) by eight years.

Classical electromagnetic concepts cannot really be applied to a classical atom since a non-quantized orbiting electron constitutes an unstable system (see below). The only situation for which such concepts can be applied meaningfully is to a free 'electron-gas', for example the electron plasma in a metal. Here again, the effect of a magnetic field is to produce diamagnetic and paramagnetic effects which exactly cancel one another. The reason is simply that in classical electromagnetism a field can deflect an electron but it cannot do work on it,* so that although the electrons are deflected into helical paths their energy distribution remains unaltered. This implies that the magnetic field produces no energy change in the system, that is the system behaves as though it has no net magnetic property. A discussion of this issue is given in a very readable and interesting form in J. H. van Vleck (1932), *The Theory of Electric and Magnetic Susceptibilities*, Oxford, Clarendon Press.

As mentioned above the classical atom is unstable; the orbiting electrons would spiral into the nucleus radiating all their energy in the process. The effective lifetime of such an atom would be infinitesimally small. Atoms exist as stable entities precisely because energy states (or angular moments) are quantized. To produce a perturbation of the energies by the classical methods is quite invalid. However, quantum mechanics shows that there is a genuine diamagnetic effect which can be calculated rigorously. The analysis is far too advanced to be given here. It gives the same answer as the classical treatment which we have given in this chapter in section 14·2. For this reason, as in most other elementary texts, we have persisted in giving the simple treatment: the answer is right though the physics is not entirely satisfactory.

* Of course a magnetic field exerts a force on a conductor through which a current is passing. If the conductor moves, work is done, but the energy does not come from the magnetic field. The energy is provided by the electric current flowing through the conductor in overcoming the back e.m.f. generated in the conductor itself as a result of its movement in the magnetic field.

269   **Magnetic properties of matter**

We have taken over from our chapter on dielectric properties the classical treatment of paramagnetism. With para-electrics quantum effects play a trivial part and the classical treatment is perfectly satisfactory. This is not so with paramagnetics. The reason is that the individual dipole can take up only a discrete number of orientations. If for example an atomic system has a resultant angular momentum represented by a vector of length $\frac{3}{2}$, the only orientations which the system can acquire in a magnetic field correspond to components of $\frac{3}{2}$, $\frac{1}{2}$, $-\frac{1}{2}$ and $-\frac{3}{2}$ in the direction of the field. In classical physics the orientations would cover the whole continuous range such that the components have values from $+\frac{3}{2}$ to $-\frac{3}{2}$. The effect on paramagnetism is not very marked but it is worth considering the simplest possible case when the dipole of the individual atom is due to a single electron spin. The only orientations allowed are those which are exactly parallel and anti-parallel to the applied field. If there are n atoms per c.c., and of these n_+ are those with orientations along the field and n_- those with orientations opposing the field, we have by the Boltzmann principle

$$n_+ = A \exp \left(\frac{pH}{kT} \right) \tag{14.42}$$

$$n_- = A \exp \left(-\frac{pH}{kT} \right). \tag{14.43}$$

Since $n_+ + n_- = n$ we find that

$$A = n \left[\exp \left(\frac{pH}{kT} \right) + \exp \left(-\frac{pH}{kT} \right) \right]^{-1}. \tag{14.44}$$

The net moment per unit volume is then

$$M = (n_+ - n_-)p \tag{14.45}$$

$$= np \frac{\left[\exp \left(\frac{pH}{kT} \right) - \exp \left(-\frac{pH}{kT} \right) \right]}{\left[\exp \left(\frac{pH}{kT} \right) + \exp \left(-\frac{pH}{kT} \right) \right]}$$

$$= np \tanh \left(\frac{pH}{kT} \right). \tag{14.46}$$

In form this curve is very similar to the Langevin function. However there is a quantitative difference in the region of small fields. Since for small values of x, $\tanh x \simeq x$ we have from equation (14.46) for small values of H,

$$M \simeq np \left(\frac{pH}{kT} \right) \simeq \frac{np^2 H}{kT}.$$

The susceptibility is then

$$\chi = \frac{M}{H} = \frac{np^2}{kT}.$$

This is just three times the value deduced from the Langevin function. Similar corrections may be applied to the discussion of ferromagnetism and the Curie temperature given above. However the correction factor of 3 will only apply rigorously for a single-electron-spin system. With most materials the situation is usually more complicated and in fact the correction factor becomes less than 3.

14·6   Ferromagnetic domains

We consider here briefly the magnetization of a single crystal of a ferro-magnetic material such as iron. With iron the 6 cube edge directions [100] are the directions of easy magnetization. If a magnetic field H is applied along the edge direction the domains are aligned easily and saturation is achieved for a very small field. If the field is applied along, say, the face diagonal direction [110] the easy orientation directions are at 45° to the field., Consequently for small fields the net magnetization in the direction of H is $1/\sqrt{2}$ of what it would be if the field were applied along the [100] direction. Experiments show that this is so (see figure 107). If the field is increased further magnetization can occur only by twisting the domain directions out of their easy orientations. This is a far more difficult process and relatively strong fields must be used to achieve it.

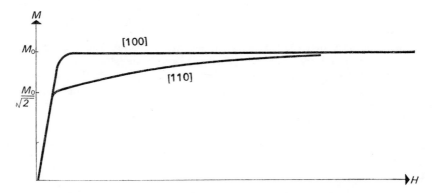

Figure 107.   Magnetization of a single crystal of iron along a cube direction [100] and along a face diagonal direction [110].

14·7   Magnetic hysteresis

When a magnetic field is applied to a magnetic material, the magnetization generally increases as the field is increased. If the field is reduced, the magnetization diminishes but does not follow the original curve – it 'lags behind'. This effect is known as magnetic hysteresis.

For the following simple description of magnetic hysteresis I am indebted to Dr Shoenberg. Consider the behaviour of a material which has a single easy

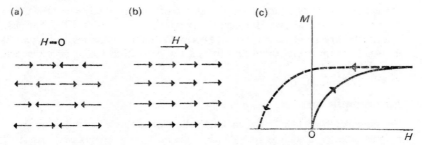

Figure 108. Magnetization of material showing a single direction of easy magnetization. (*a*) Initial unmagnetized state. (*b*) Effect of applying a weak field to the right is to orient all the domains in that direction. This gives magnetic saturation; on removing the field the magnetization remains unchanged so that the remanence is almost equal to the saturation magnetization. (*c*) Magnetization curve showing that the remanence is very large and that demagnetization can be achieved only by reversing the sign of the field.

direction of magnetization. In the unmagnetized state as many domains are magnetized to the left as to the right [see figure 108(*a*)]. If a field is applied to the right the domains all become magnetized to the right and the material is saturated [figure 108(*b*)]. The work done is used in moving the domain walls surrounding the right-pointing domains, until they swallow up the left-pointing domains. If the field is now removed no further change occurs and the material shows 100% remanence [see figure 108(*c*)]. The magnetization can be reduced only by applying a strong field to the left.

In general metal specimens contain internal stresses. If the stresses are small they merely decide which of the easy directions of magnetization are easiest of all. If the stresses are large they may completely dominate the behaviour; in that case the easy domain directions are determined by the

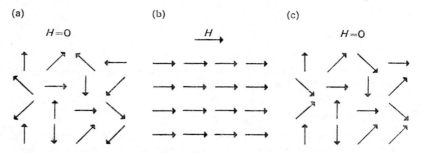

Figure 109. Magnetization of a material showing a limited number of directions of easy magnetization (vertical, horizontal and at 45°). (*a*) Initial unmagnetized state, (*b*) a horizontal field *H* aligns all the domains, (*c*) on removing the field the domains revert to their original directions but the sense is that of the field *H*. Under these conditions the remanence is about half the saturation magnetization.

272 Gases, liquids and solids

stresses rather than by crystal orientation. Consider for simplicity a specimen in which, as a result of local stresses, the domain directions are either horizontal, vertical or at 45° to these directions [figure 109(a)]. If a strong horizontal field is applied all the domains line up with the field and the material achieves saturation [figure 109(b)]. If the field is now removed the domains return to the original orientations they had in figure 109(a), but with the sense imposed on them by the applied field. This is shown in figure 109(c). Under these conditions the material has a remanence of about half the saturation magnetization.

14·8 Summary: Comparison of dielectric and magnetic properties

We now summarize in note form the main points of resemblance and difference in dielectric and magnetic properties.

Dielectric	*Magnetic*
Electronic polarization. Distortion of electron cloud. Critical frequency about 10^{15}. Induced dipole: $$p = r^3 E$$ Always in direction of E.	Diamagnetism. Deceleration or acceleration or precession of those electrons possessing orbital momentum. Susceptibility per electron: $$-\frac{1}{6}\frac{e^2 r^2}{mc^2}$$ Always opposes field H.

Unaffected by temperature.

Atomic polarization; stretching or bending of bonds if molecule has permanent dipole. Critical frequency about 10^{12}.	None.

Para-electric: orientation of molecular dipole p by applied field E, opposed by thermal motion. Typical value of $p = 5 \times 10^{-18}$ e.s.u. For small E, susceptibility per molecule is: $$\frac{p^2}{3kT}$$	Paramagnetic: orientation of atomic dipole by applied field H, opposed by thermal motion. Typical value of $p = 10^{-20}$ e.m.u. For small H, susceptibility per atom is: $$\frac{p^2}{3kT} \quad \text{(classical)}$$ $$\frac{p^2}{kT} \quad \text{(quantum mechanical)}$$

Para-electric polarization is of order 10^3 larger than paramagnetic. Hence need to consider real field rather than applied field.
Very markedly dependent on temperature.
Full behaviour described by Langevin function.

Para-electric behaviour cannot be easily enhanced by low temperatures because dielectric solidifies and molecules cannot be oriented.	Paramagnetic effects can be greatly enhanced by low temperatures since only orbits or spins have to be oriented.

Ferroelectric   *Ferromagnetic*

Co-operative phenomena in which internal field is augmented by some type of specific interaction. Large oriented domains of permanent dipoles. Above some critical temperature (Curie temperature) the interaction is ineffective, and the materials cease to be ferroelectric or ferromagnetic.

Appendixes

Appendix A

c.g.s. and MKS units

The ratio given in the right-hand column is the factor by which the quantity in the MKS system must be multiplied to convert into the c.g.s. system.

Quantity	Unit in MKS system	Unit in c.g.s. system	Ratio $\dfrac{MKS}{c.g.s.}$
Dynamical			
Length	metre m.	centimetre cm.	10^2
Mass	kilogram kg.	gram g.	10^3
Time	second sec.	second sec.	1
Density	kg. m.$^{-3}$	g. cm.$^{-3}$	10^{-3}
Velocity	m. sec.$^{-1}$	cm. sec.$^{-1}$	10^2
Acceleration	m. sec.$^{-2}$	cm. sec.$^{-2}$	10^2
Momentum	kg. m. sec.$^{-1}$	g. cm. sec.$^{-1}$	10^5
Force	newton N.	dyne	10^5
Couple	newton metre N. m.	dyne cm.	10^7
Work, energy	N. m. = Joule J.	erg	10^7
Power	watt: J. sec.$^{-1}$	erg sec.$^{-1}$	10^7
Surface tension	N. m.$^{-1}$ J. m.$^{-2}$	dyne cm.$^{-1}$; erg cm.$^{-2}$	10^3
Viscosity	kg. m.$^{-1}$ sec.$^{-1}$	g. cm.$^{-1}$ sec.$^{-1}$ (poise)	**10**

Quantity	Unit in MKS system	Unit in c.g.s. system	Ratio $\dfrac{MKS}{c.g.s.}$
Charge q	coulomb C.	e.m.u.	$\dfrac{1}{10}$
Current i	ampere A.	e.m.u.	$\dfrac{1}{10}$
Potential V	volt	e.m.u.	10^8
		e.s.u.	$\dfrac{1}{300}$
Electric field E	volt m.$^{-1}$, N.C.$^{-1}$	e.s.u.	$\frac{1}{3} \times 10^{-4}$
Electric polarization P	C. m.$^{-2}$	e.s.u.	3×10^5
Inductance L	henry	e.m.u.	10^9
Resistance R	ohm	e.m.u.	10^9
Capacitance C	farad	e.s.u.	9×10^{11}
Magnetic induction B	weber m.$^{-2}$	e.m.u.	10^4
Magnetic field H	ampere m.$^{-1}$	e.m.u.	$4\pi \times 10^3$
Intensity of mag- netization M	weber m.$^{-2}$	e.m.u.	$\dfrac{10^4}{4\pi}$
Magnetic moment m	weber m.	e.m.u.	$\dfrac{10^{10}}{4\pi}$
Volume susceptibility χ	MKS m.$^{-3}$	e.m.u. cm.$^{-3}$	$\dfrac{1}{4\pi}$

Wherever a factor 3 appears in the electrical and magnetic conversion factor, it implies the approximation for the velocity of light $c = 3 \times 10^{10}$ cm. sec. The basic definition of μ_0 in RMKS is $4\pi \times 10^{-7}$ henry m.$^{-1}$ and the permittivity is $\varepsilon_0 = (\mu_0 c^2)^{-1}$ (see B. I. Bleaney and B. Bleaney (1911), *Electricity and Magnetism*, Clarendon Press).

Appendix B
Values of some physical constants

	Usual symbol	Value c.g.s.	Value MKS
Charge on electron	e	$4 \cdot 803 \times 10^{-10}$ e.s.u. $1 \cdot 602 \times 10^{-20}$ e.m.u.	$1 \cdot 602 \times 10^{-19}$ C.
Mass of electron	m	$9 \cdot 108 \times 10^{-28}$ g.	$9 \cdot 108 \times 10^{-31}$ kg.
Electron charge: mass	$\dfrac{e}{m}$	$5 \cdot 273 \times 10^{-17}$ e.s.u. g.$^{-1}$ $1 \cdot 759 \times 10^{7}$ e.m.u. g.$^{-1}$	$1 \cdot 759 \times 10^{11}$ C. kg^{-1}.
Boltzmann constant	k	$1 \cdot 380 \times 10^{-16}$ erg deg.$^{-1}$	$1 \cdot 380 \times 10^{-23}$ joule deg.$^{-1}$
Avogadro's number	N_0	$6 \cdot 025 \times 10^{23}$ mole^{-1}	$6 \cdot 025 \times 10^{23}$ mole^{-1}
Gas constant	R	$8 \cdot 317 \times 10^{7}$ erg mole^{-1} deg.$^{-1}$ $1 \cdot 967$ cal. mole^{-1} deg.$^{-1}$	$8 \cdot 317$ joule mole^{-1} deg.$^{-1}$
Mechanical equivalent of heat	J	$4 \cdot 186 \times 10^{7}$ erg cal.$^{-1}$ $4 \cdot 186$ joule cal.$^{-1}$	— —
Vol. 1 mole gas at s.t.p.		$22 \cdot 420 \times 10^{3}$ cm.^{3}	$22 \cdot 420 \times 10^{-3}$ m.^{3}
Faraday constant	$\mathscr{F}$	$2 \cdot 894 \times 10^{-13}$ e.s.u. mole^{-1}	$96,520$ C. mole^{-1}
Planck's constant	h	$6 \cdot 625 \times 10^{-27}$ erg sec.	$6 \cdot 625 \times 10^{-34}$ joule sec.
$\left(\dfrac{h}{2\pi}\right)$	$\hbar$	$1 \cdot 054 \times 10^{-27}$ erg sec.	$1 \cdot 054 \times 10^{-34}$ joule sec.
Velocity of light	c	$2 \cdot 998 \times 10^{10}$ cm. sec.$^{-1}$	299800 km. sec.$^{-1}$

	Usual symbol	Value c.g.s.	Value MKS
Ratio proton: electron mass	$\dfrac{M}{m}$	1836	1836
Bohr magneton $\dfrac{e}{4\pi mc}$	μ_B	$9\cdot273 \times 10^{-21}$ erg gauss^{-1}	$1\cdot165 \times 10^{-29}$ weber m.$^{-1}$
Energy equiv. of 1 eV		$1\cdot602 \times 10^{-12}$ erg $23\cdot05$ kcal. mole^{-1}	$1\cdot602 \times 10^{-19}$ joule
Gravitational constant	G	$6\cdot67 \times 10^{-8}$ dyne cm.^{2} g.$^{-2}$	$6\cdot67 \times 10^{-23}$ N.m.^{2} kg.$^{-2}$
Permeability of free space	μ_0	1	$4\pi \times 10^{-7}$ henry m.$^{-1}$ (RMKS) 10^{-7} MKS
Permittivity of free space	$\varepsilon_0 = (\mu_0 c^2)^{-1}$	1	$8\cdot854\ 10^{-12}$ farad m.$^{-1}$ (RMKS) $1\cdot113\ 10^{-10}$ MKS

The values in the above table for N_0, R, J and $\mathscr{F}$ are 'physical' constants based on the number of atoms in 16 g. of ^{16}O. The equivalent constants used by chemists are sometimes based on a different gram-molecular convention and the values quoted may differ by up to one part in a thousand.

Index